McGraw-Hill's

Math

GRADE 2

New York Chicago San Francisco Lisbon London Madrid Mexico City
Milan New Delhi San Juan Seoul Singapore Sydney Toronto

The **McGraw-Hill** *Companies*

1 2 3 4 5 6 7 8 9 10 11 12 13 14 15 DOW/DOW 1 9 8 7 6 5 4 3 2

ISBN 978-0-07-177598-4
MHID 0-07-177598-6

e-ISBN 978-0-07-177599-1
e-MHID 0-07-177599-4

Cataloging-in-Publication data for this title are on file at the Library of Congress.

Printed and bound by RR Donnelley.

Editorial Services: Pencil Cup Press
Production Services: Jouve
Illustrator: Eileen Hine
Designer: Ella Hanna

This book is printed on acid-free paper.

Table of Contents

Table of Contents

Welcome to McGraw-Hill's Math!

This book will help you do well in mathematics.
Its lessons explain the math.
Then they provide practice activities.

Open your book. Look at the Table of Contents.
It tells what topics are covered in each lesson.

Then look at the 10-Week Summer Study Plan.
It shows one way to plan your time.
You may be able to do some lessons more quickly.
You may need to take more time to do other lessons.

Each group of lessons ends with a Chapter Test.
The Chapter Tests will show you what skills you have learned.
It will show you what skills you may need to practice more.

There are 2 Reviews in this book.
One is after Chapter 4. One is after Chapter 9.
Complete these pages. They will show you how much you have learned.

10-Week Summer Study Plan

Many children will use this book as a summer study program.
Use this 10-week study plan to help plan the time.
Put a ✔ in the box when the child finishes the day's work.

	Day	Lesson Pages	Test Pages
Week 1	Monday	8, 9	
	Tuesday	10	
	Wednesday	11, 12	
	Thursday	13, 14	
	Friday	15	16–17
Week 2	Monday	18, 19	
	Tuesday	20, 21	
	Wednesday	22, 23	
	Thursday	24, 25	
	Friday	26	27–28
Week 3	Monday	29, 30	
	Tuesday	31, 32	
	Wednesday	33, 34	35–36
	Thursday	37, 38	
	Friday	39, 40, 41	
Week 4	Monday	42	
	Tuesday	43	
	Wednesday	44	
	Thursday	45	46–47
	Friday		REVIEW 48–53
Week 5	Monday	54	
	Tuesday	55	
	Wednesday	56, 57	
	Thursday	58	
	Friday	59	

10-Week Summer Study Plan

	Day	Lesson Pages	Test Pages
Week 6	Monday	60, 61	
	Tuesday	62	
	Wednesday	63	64–65
	Thursday	66, 67	
	Friday	68, 69	
Week 7	Monday	70, 71	
	Tuesday	72, 73	
	Wednesday	74	75–76
	Thursday	77, 78	
	Friday	79, 80	
Week 8	Monday	81, 82	
	Tuesday	83, 84	
	Wednesday	85	86–87
	Thursday	88, 89	
	Friday	90, 91	
Week 9	Monday	92	
	Tuesday	93	
	Wednesday	94, 95	
	Thursday	96	
	Friday	97	98–99
Week 10	Monday	100, 101	
	Tuesday	102, 103	
	Wednesday	104, 105	
	Thursday		106–107
	Friday		REVIEW 108–113

Name ______________________

Adding Through 20

When you add, you put groups together.
Then you tell how many in all. This is called the sum.

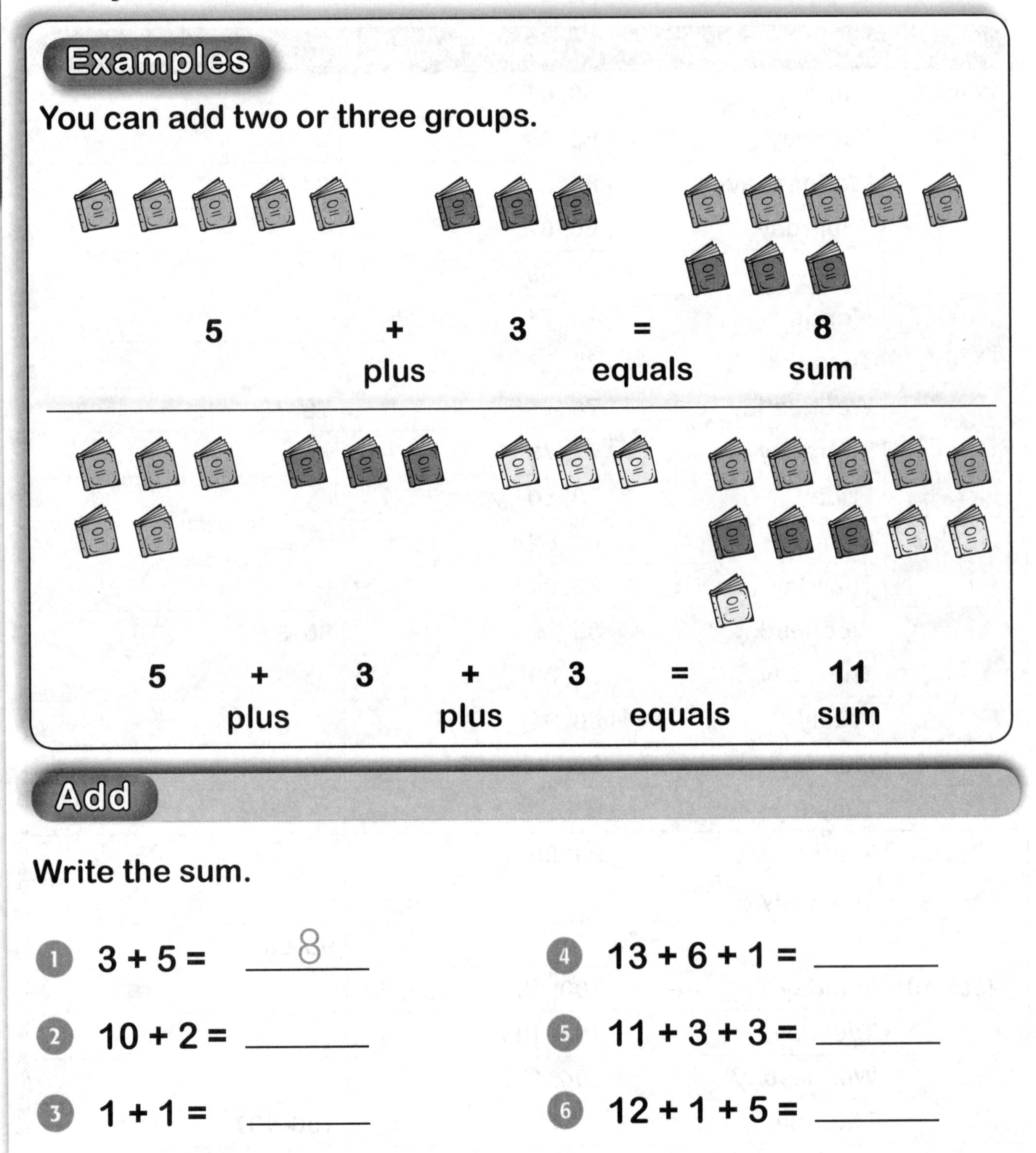

Examples

You can add two or three groups.

5 + 3 = 8

(+ plus, = equals, 8 sum)

5 + 3 + 3 = 11

(+ plus, + plus, = equals, 11 sum)

Add

Write the sum.

1. $3 + 5 =$ 8
2. $10 + 2 =$ ______
3. $1 + 1 =$ ______
4. $13 + 6 + 1 =$ ______
5. $11 + 3 + 3 =$ ______
6. $12 + 1 + 5 =$ ______

Name ________________________

Subtracting Through 20

You subtract to tell how many are left.
This is called the difference.

Example

12 – 4 = 8

minus, equals, difference

Subtract

Write the difference.

1. 20 – 5 = 15
2. 10 – 2 = ______
3. 5 – 2 = ______
4. 9 – 8 = ______
5. 15 – 4 = ______
6. 16 – 6 = ______
7. 7 – 5 = ______
8. 4 – 4 = ______
9. 17 – 5 = ______
10. 11 – 3 = ______
11. 10 – 8 = ______
12. 12 – 9 = ______
13. 8 – 0 = ______
14. 8 – 2 = ______

Name ______________________________

More Adding and Subtracting Through 20

You add to tell how many in all.
You subtract to tell how many are left.
When you know an addition fact, you know a subtraction fact too.

Example

6 + 5 = 11

11 − 5 = 6

Add or Subtract

Write the sum or difference.

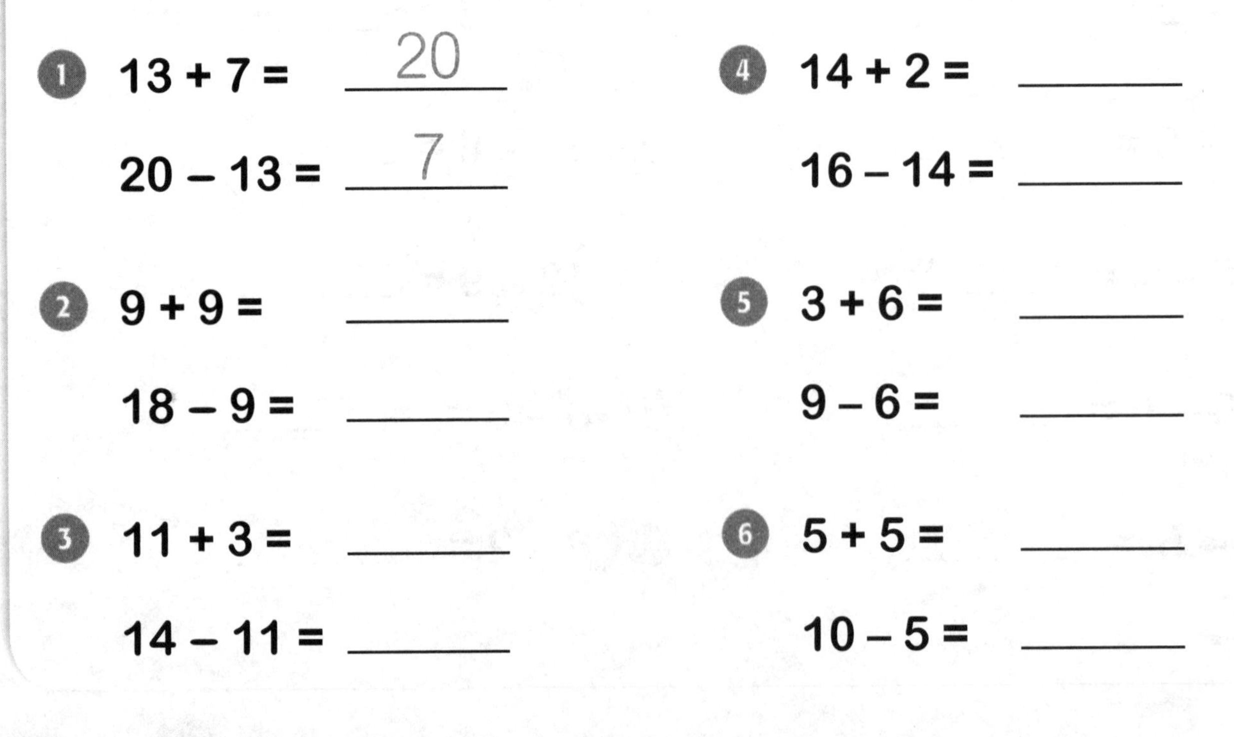

1. 13 + 7 = 20
   20 − 13 = 7

2. 9 + 9 = ______
   18 − 9 = ______

3. 11 + 3 = ______
   14 − 11 = ______

4. 14 + 2 = ______
   16 − 14 = ______

5. 3 + 6 = ______
   9 − 6 = ______

6. 5 + 5 = ______
   10 − 5 = ______

Name ______________________

One-Step Addition Word Problems

You can solve a word problem by adding.

Example

Jane sees 2 red birds.
Then she sees 3 blue birds.
How many birds does Jane see in all?

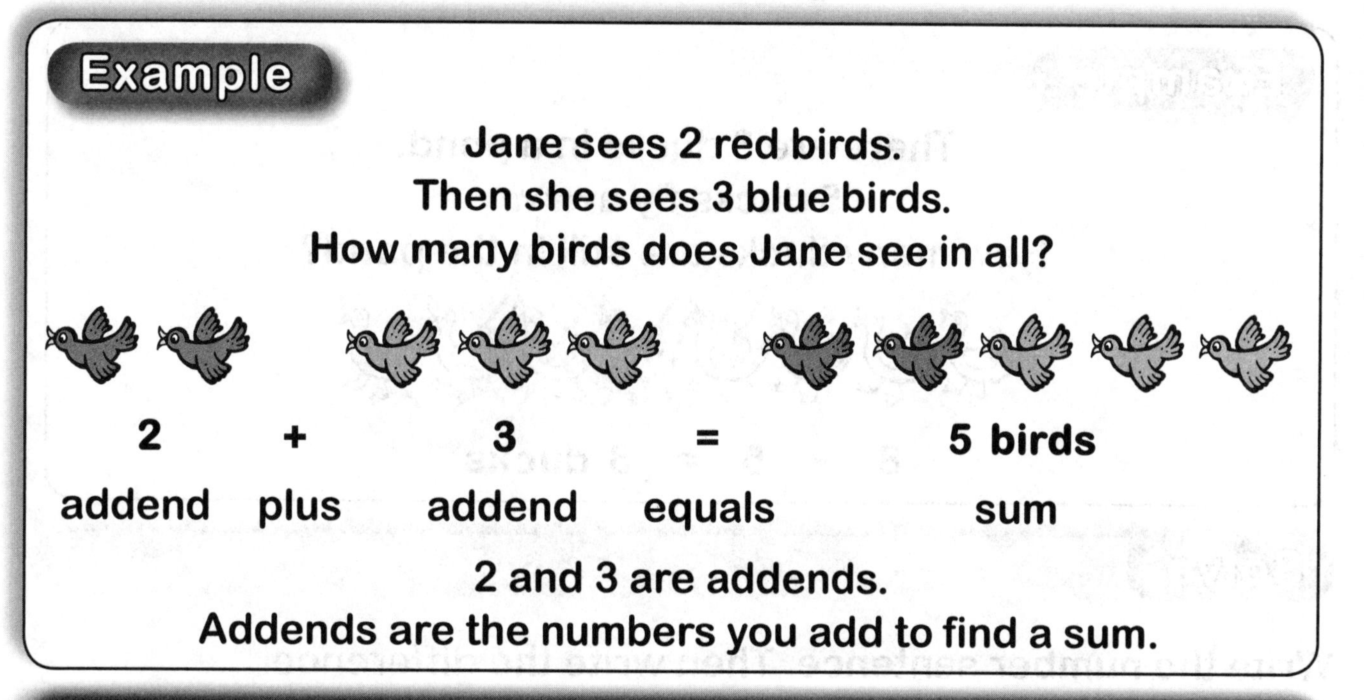

2	+	3	=	5 birds
addend	plus	addend	equals	sum

2 and 3 are addends.
Addends are the numbers you add to find a sum.

Solve

Write the addends. Then write the sum.

1. Sam saw 4 fish.
   Then he saw 2 more fish.
   How many fish did Sam see in all?

   4 + 2 = 6 fish

2. Jose has 12 books. Maria has 6 books. How many books do they have in all?

   ____ + ____ = ____ books

3. Dee has 3 toy cars. Her brother has 3 toy cars. How many toy cars do they have in all?

   ____ + ____ = ____ toy cars

Name ______________________________

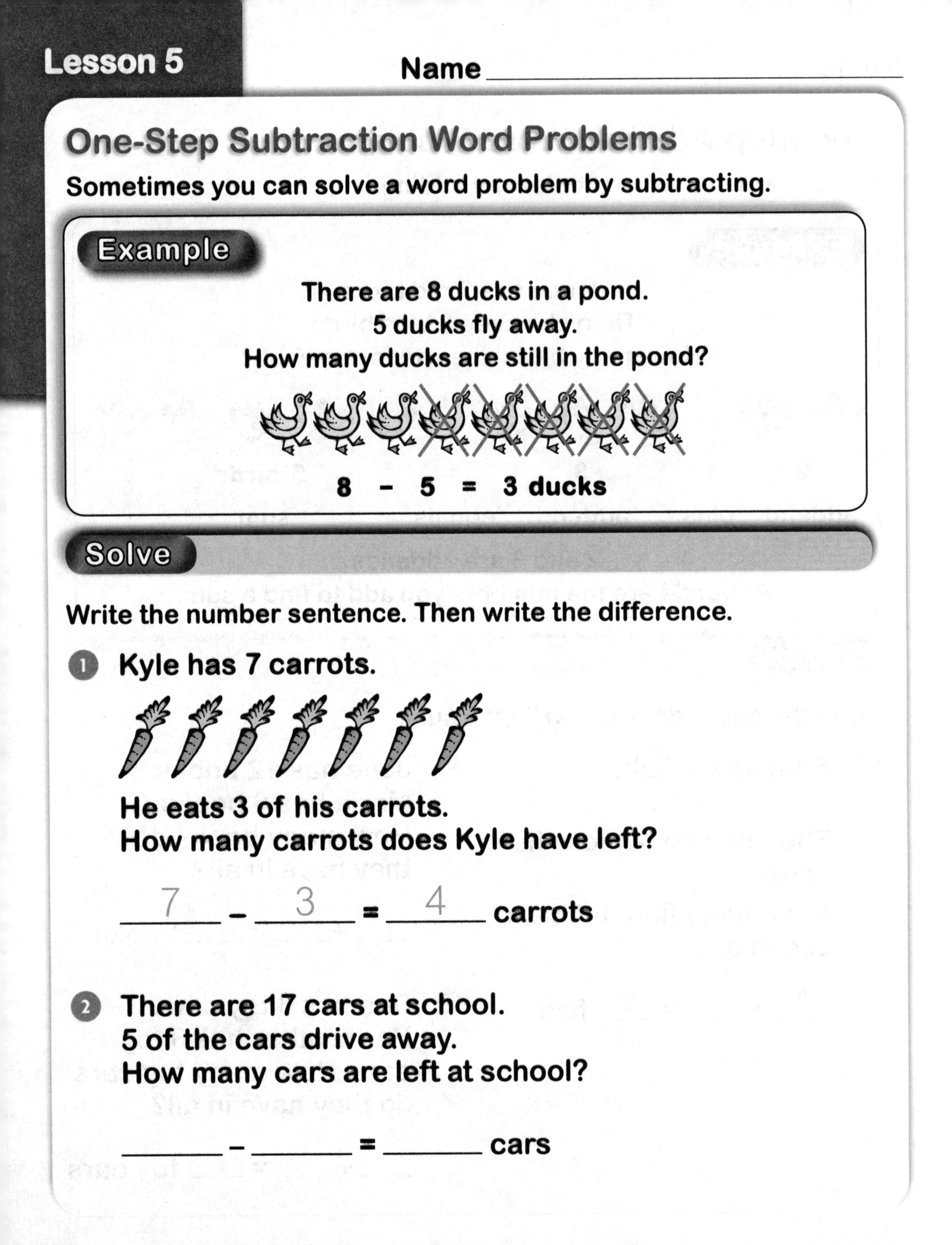

One-Step Subtraction Word Problems

Sometimes you can solve a word problem by subtracting.

Example

There are 8 ducks in a pond.
5 ducks fly away.
How many ducks are still in the pond?

8 - 5 = 3 ducks

Solve

Write the number sentence. Then write the difference.

1. Kyle has 7 carrots.

He eats 3 of his carrots.
How many carrots does Kyle have left?

__7__ - __3__ = __4__ carrots

2. There are 17 cars at school.
5 of the cars drive away.
How many cars are left at school?

______ - ______ = ______ cars

Name ______________________

Two-Step Word Problems

Some problems take two steps to solve. You may have to subtract first. Then you use your answer to add. The sum is the answer to the problem.

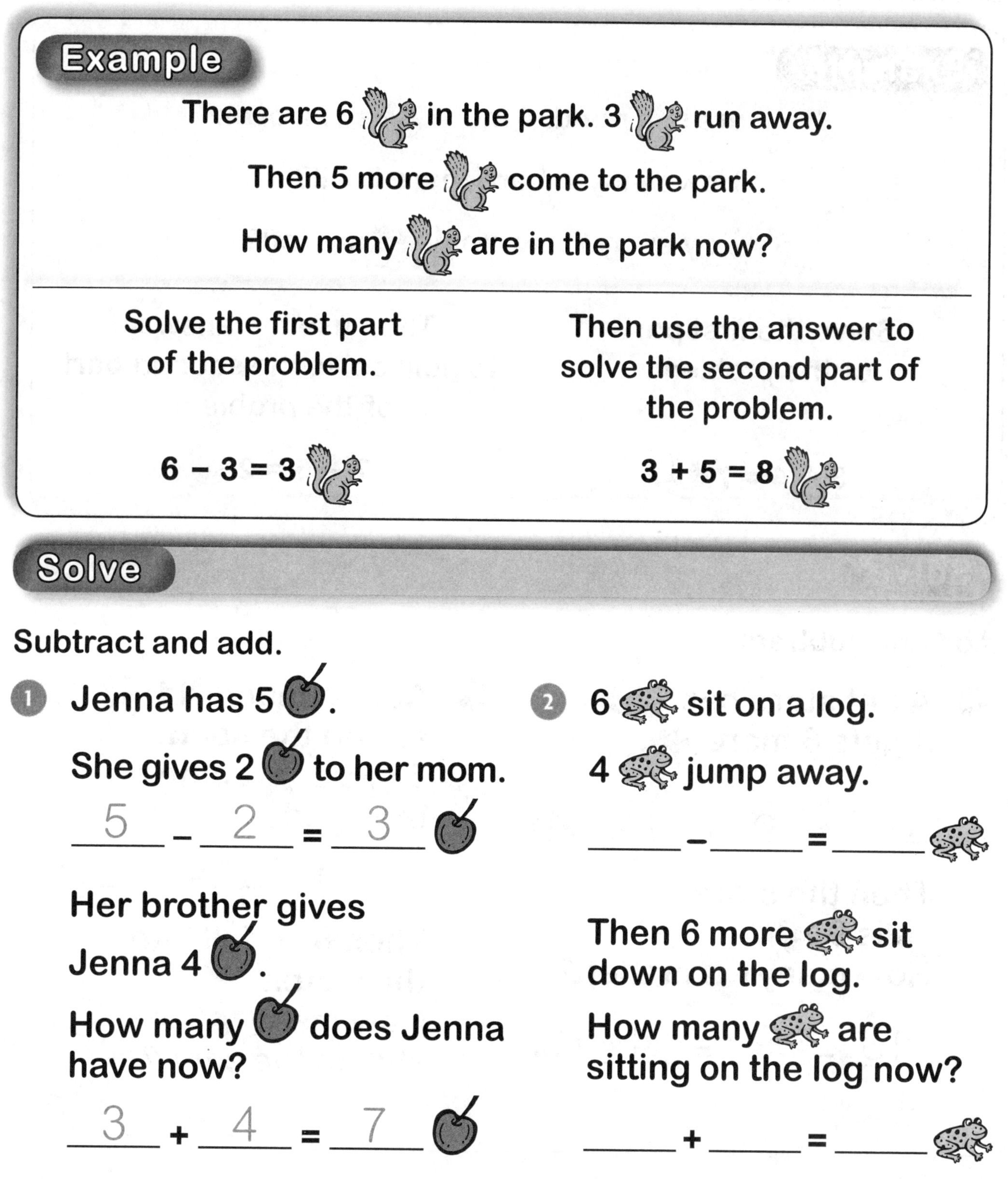

Example

There are 6 in the park. 3 run away.

Then 5 more come to the park.

How many are in the park now?

Solve the first part of the problem.

$6 - 3 = 3$

Then use the answer to solve the second part of the problem.

$3 + 5 = 8$

Solve

Subtract and add.

1. Jenna has 5.

   She gives 2 to her mom.

   5 – 2 = 3

   Her brother gives Jenna 4.

   How many does Jenna have now?

   3 + 4 = 7

2. 6 sit on a log.

   4 jump away.

   ______ – ______ = ______

   Then 6 more sit down on the log.

   How many are sitting on the log now?

   ______ + ______ = ______

Lesson 7

Name ____________________

More Two-Step Word Problems

Some problems take two steps to solve. You may have to add first. Then you use your answer to subtract. The difference is the answer to the problem.

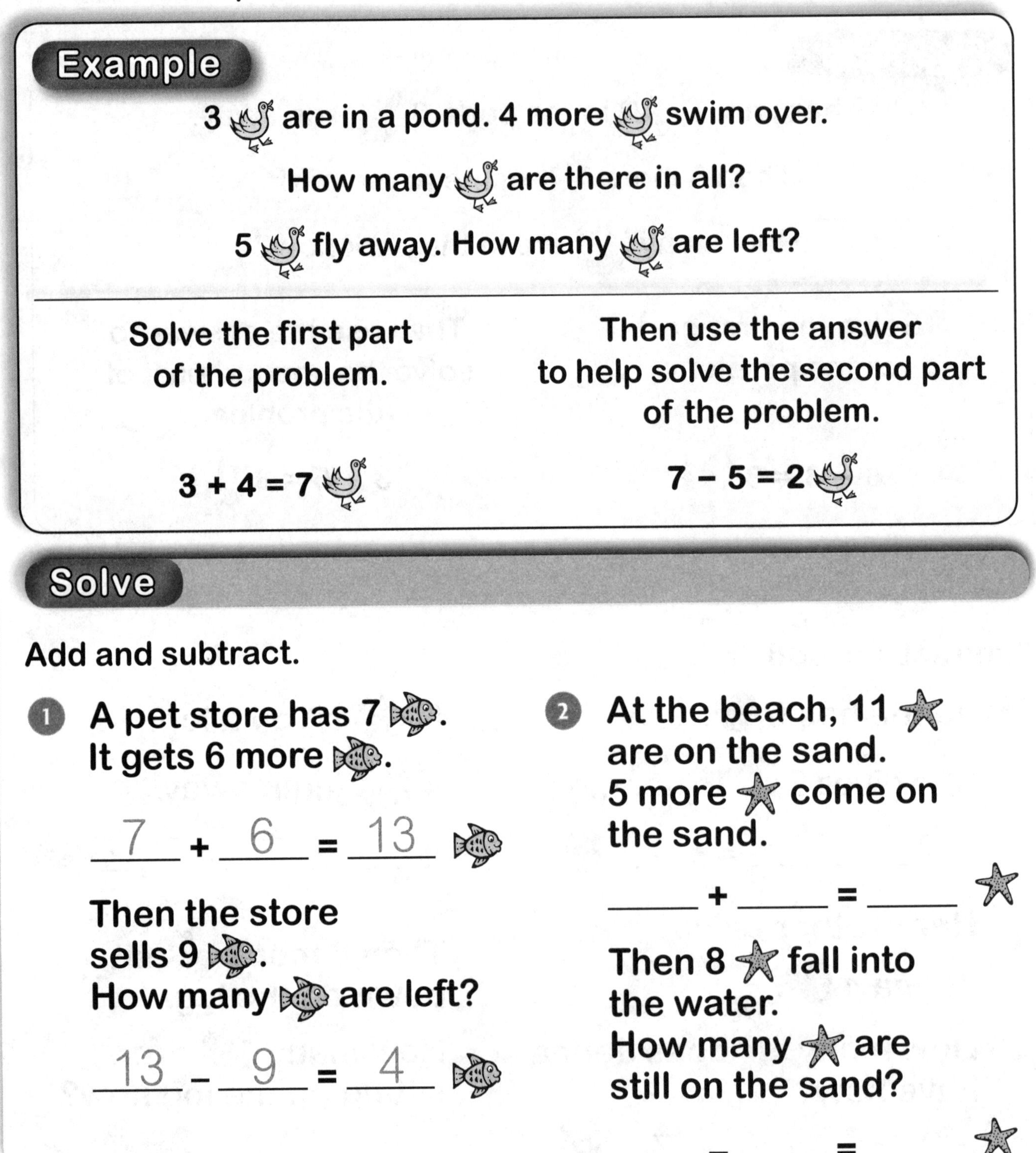

Example

3 ducks are in a pond. 4 more ducks swim over.

How many ducks are there in all?

5 ducks fly away. How many ducks are left?

Solve the first part of the problem.

$3 + 4 = 7$ ducks

Then use the answer to help solve the second part of the problem.

$7 - 5 = 2$ ducks

Solve

Add and subtract.

1. A pet store has 7 fish. It gets 6 more fish.

   __7__ + __6__ = __13__ fish

   Then the store sells 9 fish. How many fish are left?

   __13__ – __9__ = __4__ fish

2. At the beach, 11 starfish are on the sand. 5 more starfish come on the sand.

   ____ + ____ = ____ starfish

   Then 8 starfish fall into the water. How many starfish are still on the sand?

   ____ – ____ = ____ starfish

Name ______________________________

Pictures and Number Sentences

You can draw a picture and write a number sentence to show a problem. Then you solve the problem.

Example

7 squirrels are on a rock.
4 of them run away.
How many squirrels are still on the rock?

Step 1: Draw a picture to show the problem.

Step 2: Write a number sentence for the problem.

$7 - 4 = ?$

Step 3: Solve.

$7 - 4 = 3$

Solve

Draw a picture and write a number sentence to solve.

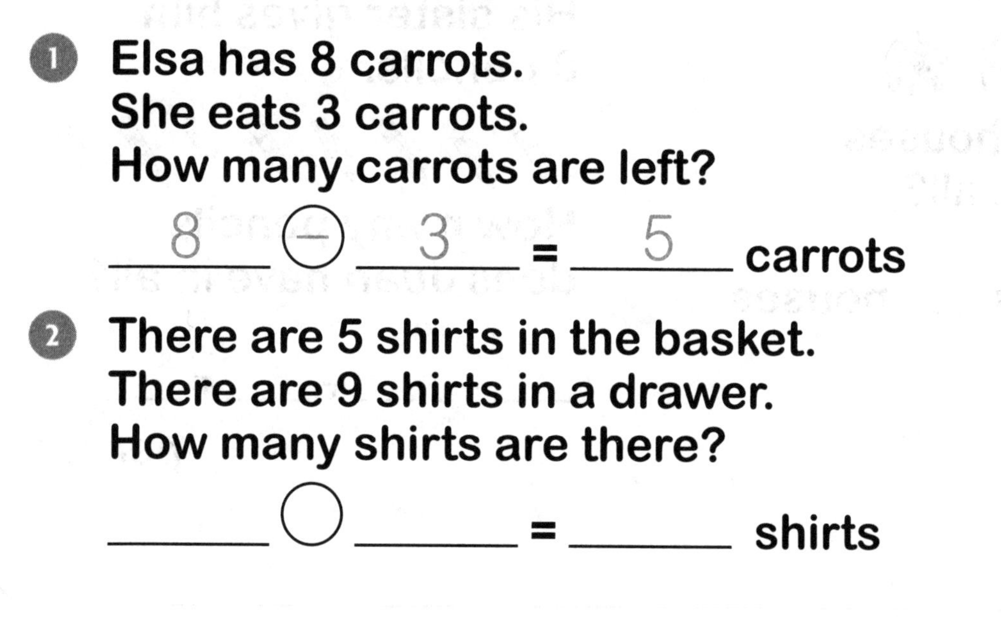

1. Elsa has 8 carrots.
   She eats 3 carrots.
   How many carrots are left?

   __8__ (−) __3__ = __5__ carrots

2. There are 5 shirts in the basket.
   There are 9 shirts in a drawer.
   How many shirts are there?

   _____ ◯ _____ = _____ shirts

Name ________________________

Add or subtract. Write the sum or difference.

1. $2 + 4 =$ ______
2. $13 + 4 =$ ______
3. $6 + 7 + 6 =$ ______
4. $7 + 1 + 4 =$ ______
5. $20 - 7 =$ ______
6. $6 - 5 =$ ______
7. $14 - 14 =$ ______
8. $19 - 5 =$ ______

Write the number sentence. Add or subtract to solve.

9. There are 6 houses on the right side of the street.

There are 4 houses on the left.

How many houses are there in all?

____ + ____ = ____ houses

10. Juan has 5 pencils.

His friend gives him 6 pencils.

His sister gives him 8 pencils.

How many pencils does Juan have in all?

____ + ____ + ____ = ____ pencils

Name ______________________

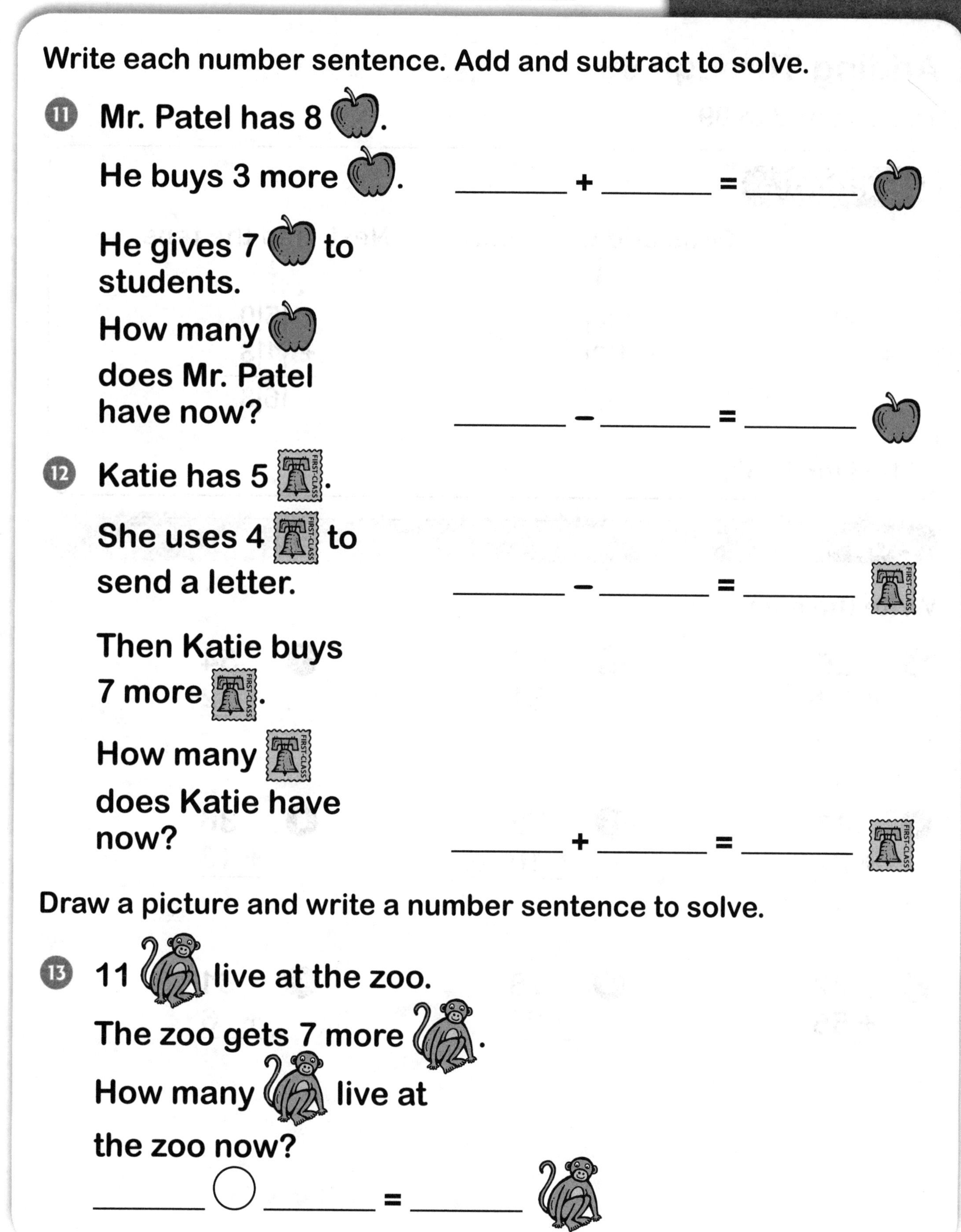

Write each number sentence. Add and subtract to solve.

11 Mr. Patel has 8 .

He buys 3 more .

_____ + _____ = _____

He gives 7 to students.

How many does Mr. Patel have now?

_____ – _____ = _____

12 Katie has 5 .

She uses 4 to send a letter.

_____ – _____ = _____

Then Katie buys 7 more .

How many does Katie have now?

_____ + _____ = _____

Draw a picture and write a number sentence to solve.

13 11 live at the zoo.

The zoo gets 7 more .

How many live at the zoo now?

_____ ◯ _____ = _____

Name ________________

Adding Through 99

You can add to 99.

Example

First, add the ones. Next, add the tens.

	First, add the ones.	Next, add the tens.
50	5|0|	|5|0
+ 13	+ 1|3|	+ |1|3
	|3|	|6|3

The sum is 63.

Add

Write the sum.

1. 25 + 33 = 58
2. 17 + 81
3. 54 + 12
4. 47 + 21
5. 72 + 10
6. 36 + 13
7. 22 + 55
8. 78 + 21
9. 61 + 6

Name ______________________________

Lesson 2

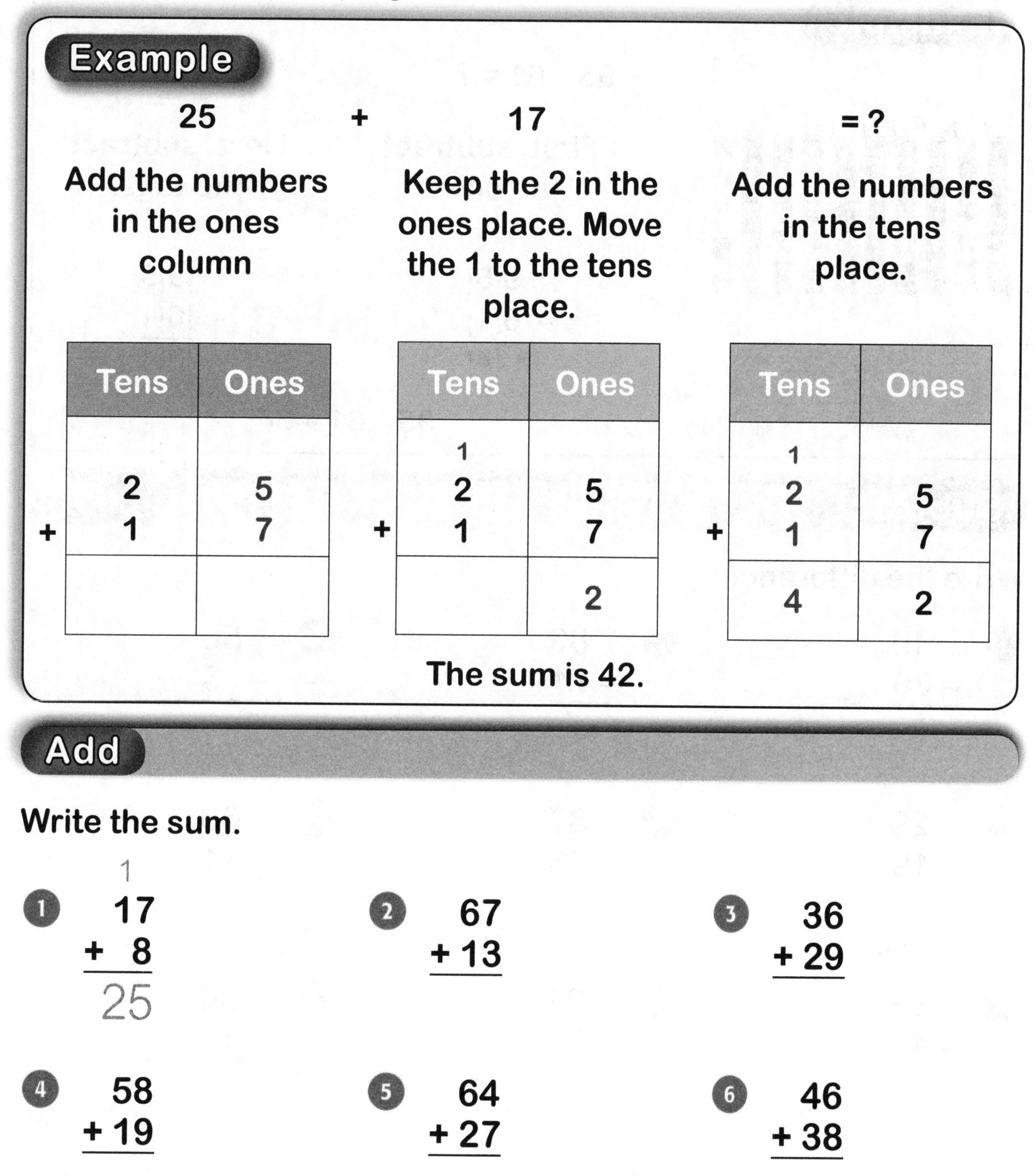

Adding Through 99 with Regrouping

Sometimes, adding the ones forms another group of 10. This is called regrouping.

Example

25 + 17 = ?

Add the numbers in the ones column

	Tens	Ones
	2	5
+	1	7

Keep the 2 in the ones place. Move the 1 to the tens place.

	Tens	Ones
	1	
	2	5
+	1	7
		2

Add the numbers in the tens place.

	Tens	Ones
	1	
	2	5
+	1	7
	4	2

The sum is 42.

Add

Write the sum.

1. $\begin{array}{r} ^{1}\\ 17 \\ +\ \ 8 \\ \hline 25 \end{array}$

2. $\begin{array}{r} 67 \\ +13 \\ \hline \end{array}$

3. $\begin{array}{r} 36 \\ +29 \\ \hline \end{array}$

4. $\begin{array}{r} 58 \\ +19 \\ \hline \end{array}$

5. $\begin{array}{r} 64 \\ +27 \\ \hline \end{array}$

6. $\begin{array}{r} 46 \\ +38 \\ \hline \end{array}$

Name ____________________

Subtracting Through 99

When you subtract, you find the difference.

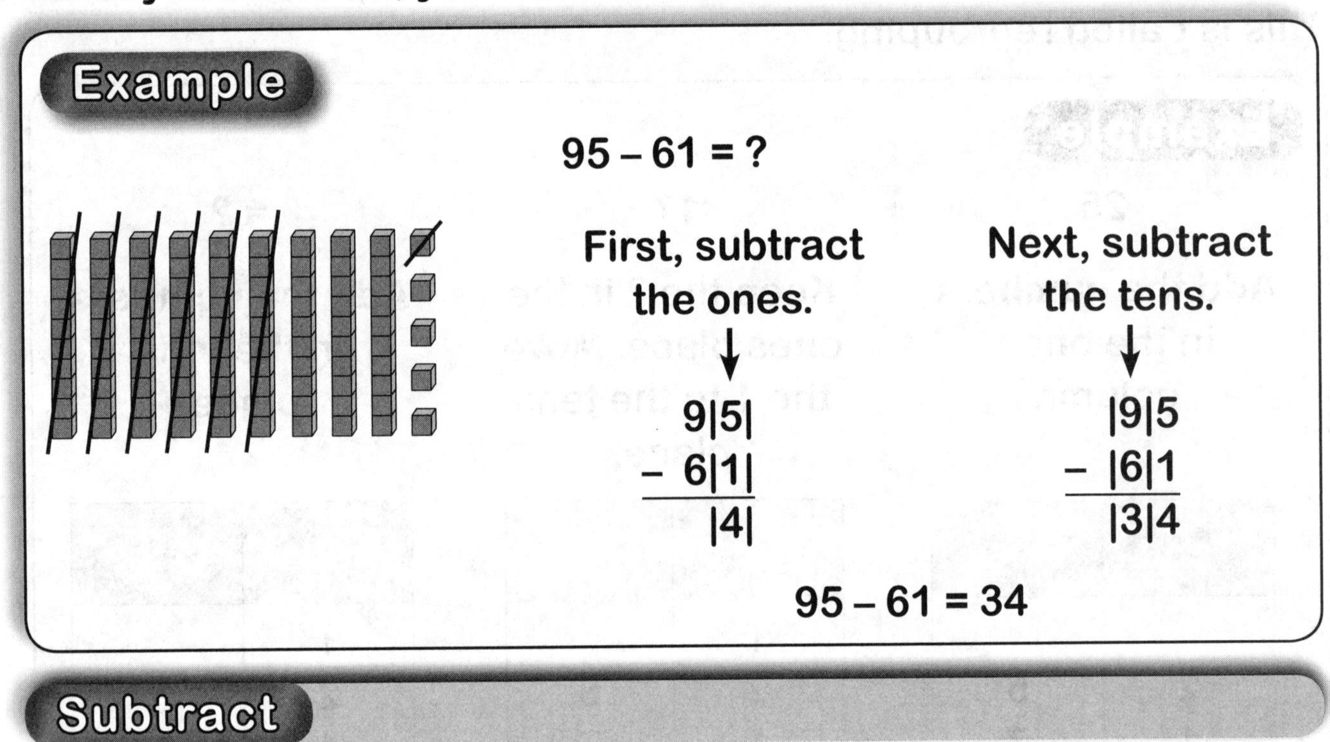

Subtract

Write the difference.

1. $\begin{array}{r} 49 \\ -26 \\ \hline 23 \end{array}$

2. $\begin{array}{r} 83 \\ -71 \\ \hline \end{array}$

3. $\begin{array}{r} 56 \\ -\ \ 5 \\ \hline \end{array}$

4. $\begin{array}{r} 29 \\ -19 \\ \hline \end{array}$

5. $\begin{array}{r} 37 \\ -26 \\ \hline \end{array}$

6. $\begin{array}{r} 68 \\ -51 \\ \hline \end{array}$

7. $\begin{array}{r} 53 \\ -41 \\ \hline \end{array}$

8. $\begin{array}{r} 72 \\ -30 \\ \hline \end{array}$

9. $\begin{array}{r} 64 \\ -11 \\ \hline \end{array}$

Name ______________________

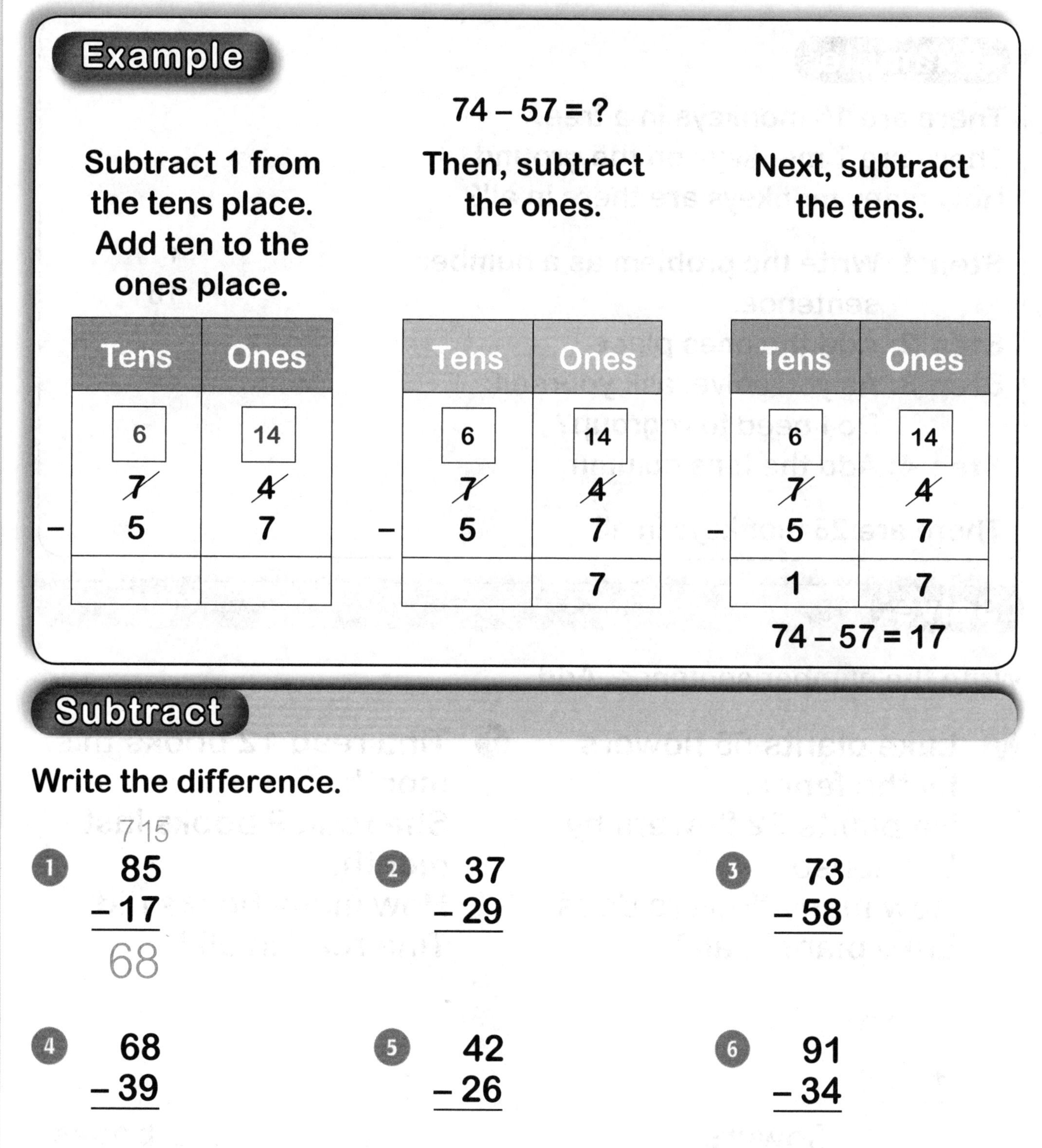

Subtracting Through 99 with Regrouping

Sometimes, there are not enough ones to subtract.

Example

74 – 57 = ?

Subtract 1 from the tens place. Add ten to the ones place.

	Tens	Ones
	6	14
	~~7~~	~~4~~
–	5	7

Then, subtract the ones.

	Tens	Ones
	6	14
	~~7~~	~~4~~
–	5	7
		7

Next, subtract the tens.

	Tens	Ones
	6	14
	~~7~~	~~4~~
–	5	7
	1	7

74 – 57 = 17

Subtract

Write the difference.

1. 7 15
85
– 17
68

2. 37
– 29

3. 73
– 58

4. 68
– 39

5. 42
– 26

6. 91
– 34

Name ______________________

One-Step Addition Word Problems

You can solve some word problems by addition.

Example

There are 16 monkeys in a tree.
There are 7 monkeys on the ground.
How many monkeys are there in all?

Step 1: Write the problem as a number sentence.
Step 2: Add the ones place.
Step 3: As you solve, ask yourself: Do I need to regroup?
Step 4: Add the tens column.

$$\begin{array}{r} {}^{1} \\ 16 \\ +\ \ 7 \\ \hline 23 \end{array}$$

There are 23 monkeys in all.

Solve

Write the number sentence. Add.

1. Luke plants 55 flowers by the fence. He plants 32 flowers by the house. How many flowers does Luke plant in all?

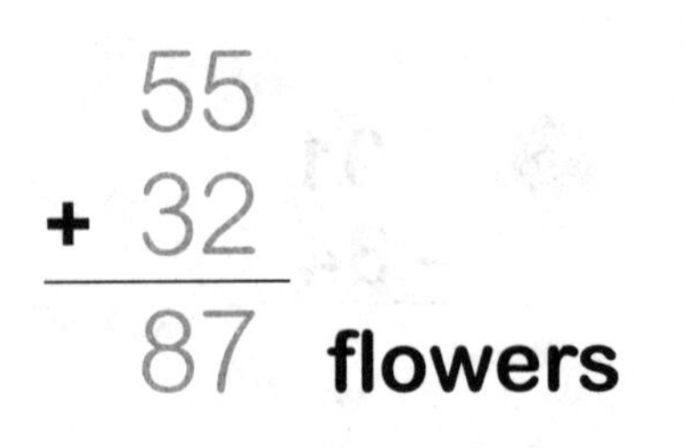

2. Tina read 12 books this month. She read 9 books last month. How many books did Tina read in all?

______ books

Name ____________________

One-Step Subtraction Word Problems

You can solve some word problems by subtracting.

Example

Kerri has 28 baseball cards.
She gives 19 to her brother.
How many baseball cards does Kerri have left?

Step 1: Write the problem as a number sentence.
Step 2: Ask yourself: Do I need to regroup?
Step 3: Regroup if necessary.
Step 4: Subtract the ones.
Step 5: Subtract the tens.

$$\begin{array}{r} \scriptstyle 1\ 18 \\ \not{2}\not{8} \\ -\ 19 \\ \hline 9 \end{array}$$

Kerri has 9 baseball cards left.

Solve

Write the number sentence. Subtract.

1. There are 37 toys in Chad's room.
There are 25 toys in Eva's room.
How many more toys are in Chad's room?

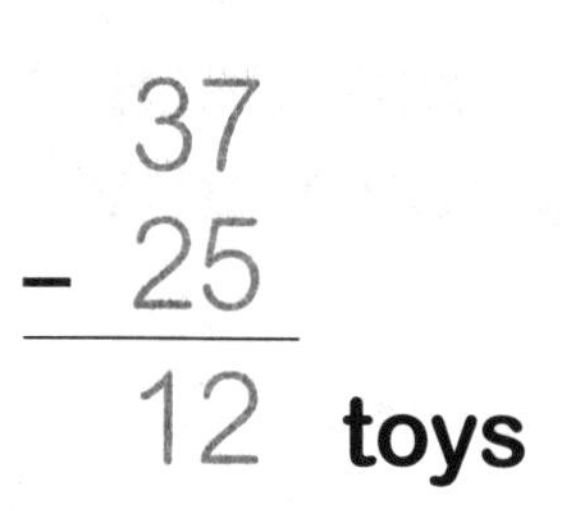

toys

2. Mona's bag has 77 marbles. She gives away 49 marbles.
How many marbles are left in the bag?

______ marbles

Lesson 7

Name ______________________

Two-Step Word Problems

Some problems take two steps to solve. You may have to subtract first. Then you use your answer and add. The sum is the answer to the problem.

Example

Lamar has 18 books.
He gives 4 books to his brother.
Then his dad gives Lamar 8 more books.
How many books does Lamar have now?

Solve the first part of the problem.

18 – 4 = 14 books

Then use the answer to help solve the second part of the problem.

14 + 8 = 22 books

Lamar now has 22 books.

Solve

Write the number sentence for both steps. Subtract and add to solve.

1. 38 cars are in the parking lot. 12 cars leave. Then 25 more cars arrive. How many cars are in the lot now?

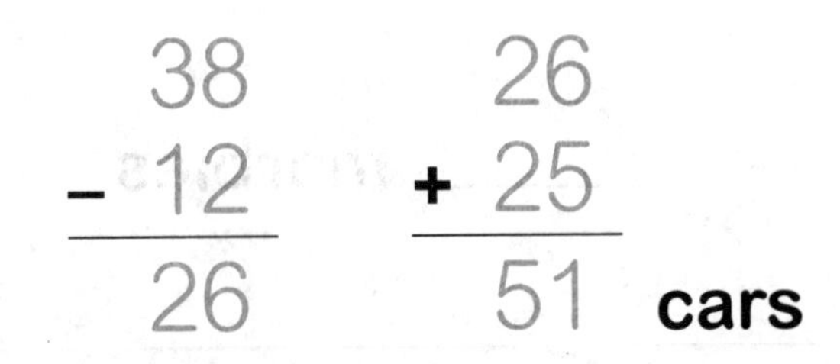

$$\begin{array}{r} 38 \\ -\ 12 \\ \hline 26 \end{array} \qquad \begin{array}{r} 26 \\ +\ 25 \\ \hline 51 \end{array}$$

cars

2. Mr. Jones has 71 markers. He gives 46 to his students. Then he gets 20 more markers. How many markers does Mr. Jones have now?

______ markers

Name ______________________________

More Two-Step Word Problems

Some problems take two steps to solve. You may have to add first. Then you use your answer and subtract. The difference is the answer to the problem.

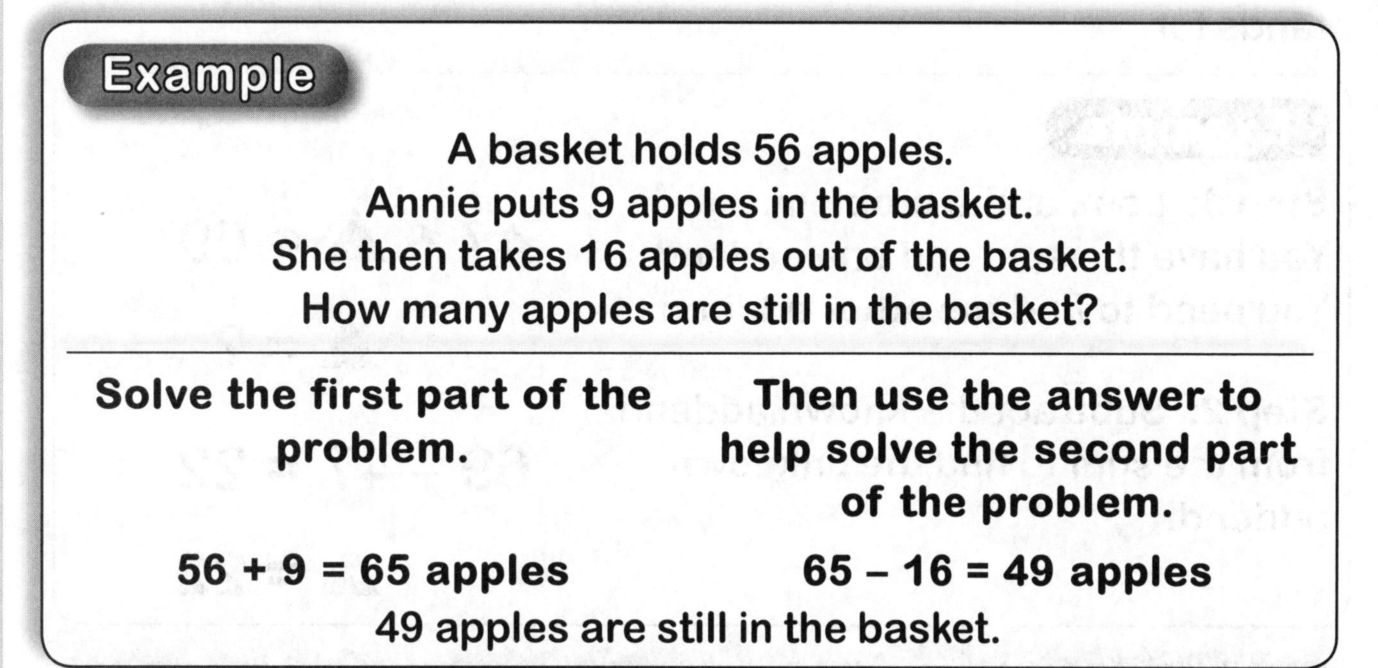

Example

A basket holds 56 apples.
Annie puts 9 apples in the basket.
She then takes 16 apples out of the basket.
How many apples are still in the basket?

Solve the first part of the problem.	**Then use the answer to help solve the second part of the problem.**
56 + 9 = 65 apples	**65 – 16 = 49 apples**

49 apples are still in the basket.

Solve

Write the number sentence for both steps. Add and subtract to solve.

1. Carl makes 35 blue hats.
   Then he makes 52 red hats.
   He sells 46 hats.
   How many hats does Carl have left?

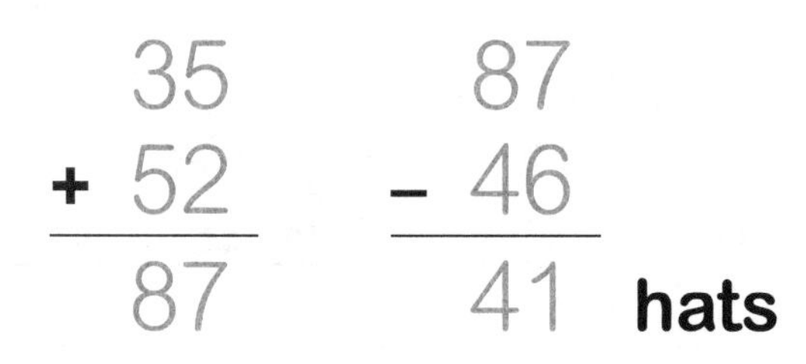

hats

2. Jan has 39 stamps with flags.
   She has 43 stamps with flowers.
   She uses 58 stamps.
   How many stamps does Jan have left?

_______ stamps

Name ______________________

Symbols for Unknown Numbers

A symbol stands for something else. You can replace the symbol in a math problem with a number. Where is the symbol? Decide if you need to add or subtract to find the number the symbol stands for.

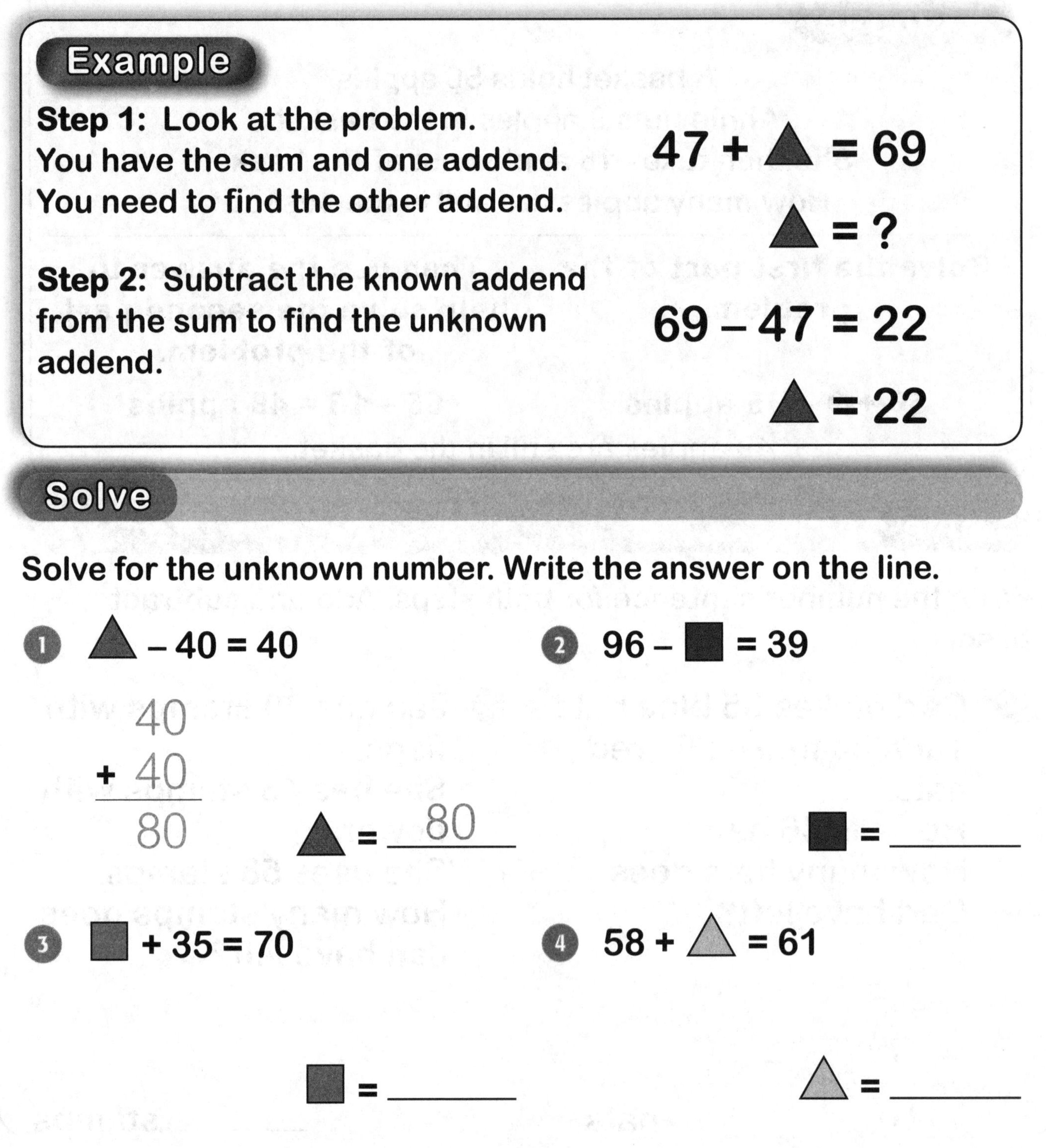

Example

Step 1: Look at the problem. You have the sum and one addend. You need to find the other addend.

47 + ▲ = 69

▲ = ?

Step 2: Subtract the known addend from the sum to find the unknown addend.

69 – 47 = 22

▲ = 22

Solve

Solve for the unknown number. Write the answer on the line.

1. ▲ – 40 = 40

   40
   + 40
   80

   ▲ = 80

2. 96 – ■ = 39

   ■ = ______

3. ■ + 35 = 70

   ■ = ______

4. 58 + ▲ = 61

   ▲ = ______

Name ______________________________

Add. Write the sum.

1. $\begin{array}{r} 84 \\ +\ 13 \\ \hline \end{array}$

2. $\begin{array}{r} 71 \\ +\ 17 \\ \hline \end{array}$

3. $\begin{array}{r} 12 \\ +\ 87 \\ \hline \end{array}$

4. $\begin{array}{r} 36 \\ +\ 40 \\ \hline \end{array}$

5. $\begin{array}{r} 37 \\ +\ 37 \\ \hline \end{array}$

6. $\begin{array}{r} 39 \\ +\ 45 \\ \hline \end{array}$

7. $\begin{array}{r} 78 \\ +\ 13 \\ \hline \end{array}$

8. $\begin{array}{r} 58 \\ +\ \ 5 \\ \hline \end{array}$

Subtract. Write the difference.

9. $\begin{array}{r} 97 \\ -\ 16 \\ \hline \end{array}$

10. $\begin{array}{r} 34 \\ -\ 34 \\ \hline \end{array}$

11. $\begin{array}{r} 58 \\ -\ 35 \\ \hline \end{array}$

12. $\begin{array}{r} 76 \\ -\ 42 \\ \hline \end{array}$

13. $\begin{array}{r} 54 \\ -\ 47 \\ \hline \end{array}$

14. $\begin{array}{r} 85 \\ -\ 77 \\ \hline \end{array}$

15. $\begin{array}{r} 65 \\ -\ 39 \\ \hline \end{array}$

16. $\begin{array}{r} 71 \\ -\ \ 8 \\ \hline \end{array}$

Add or subtract. Write the sum or difference.

17. There are 58 people on a plane.
Then 21 more people get on the plane.
How many people are on the plane now?

____ + ____ = ____ people

18. One box has 48 crayons.
Another box has 24 crayons.
How many more crayons are in the box of 48?

____ – ____ = ____ crayons

Chapter 2 Test

Name ___________________________

Write the number sentence for both steps. Add and subtract to solve. Write the answer.

19 Tara has 20 stickers.
She gives 18 stickers away.
Then Tara gets 12 more stickers.
How many stickers does Tara have now?

_____ – _____ = _____

___ + ___ = ___ stickers

20 There are 26 books on one shelf in the classroom.
There are 38 books on another shelf.
The students take 20 of the books to read.
How many books are still on the shelves?

_____ + _____ = _____

___ – ___ = ___ books

Solve for the unknown number. Write a number sentence. Write the answer.

21 58 + ▲ = 94

▲ = _______

22 ■ – 27 = 33

■ = _______

23 89 – ■ = 53

■ = _______

24 ▲ + 74 = 99

▲ = _______

Name ______________________________

Even and Odd Numbers

Numbers can be even or odd.

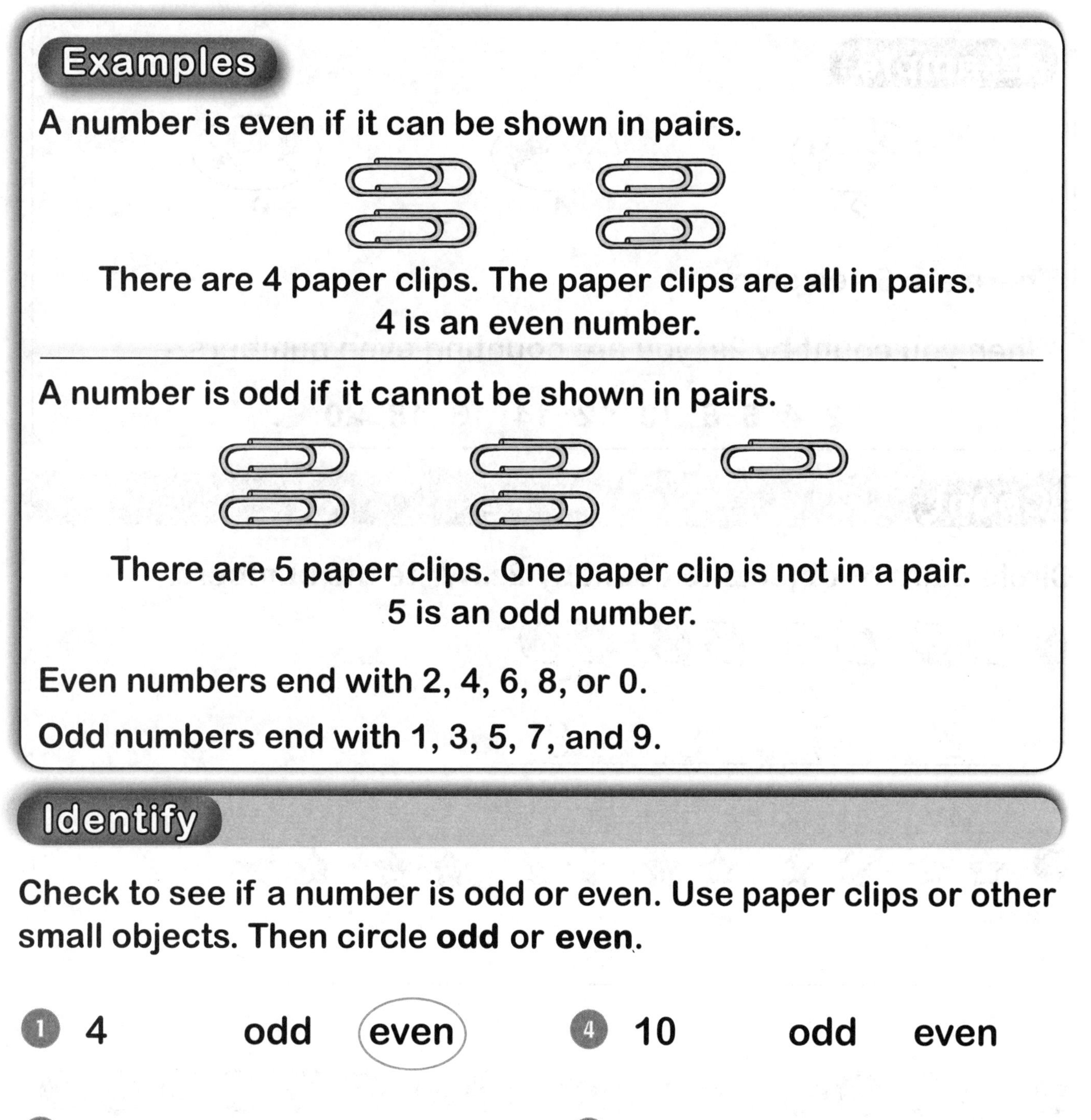

Examples

A number is even if it can be shown in pairs.

There are 4 paper clips. The paper clips are all in pairs.
4 is an even number.

A number is odd if it cannot be shown in pairs.

There are 5 paper clips. One paper clip is not in a pair.
5 is an odd number.

Even numbers end with 2, 4, 6, 8, or 0.

Odd numbers end with 1, 3, 5, 7, and 9.

Identify

Check to see if a number is odd or even. Use paper clips or other small objects. Then circle **odd** or **even**.

1. 4 odd (even)
2. 13 odd even
3. 7 odd even
4. 10 odd even
5. 8 odd even
6. 19 odd even

Name ____________________

Counting by 2s

You can pair objects to count by 2s.

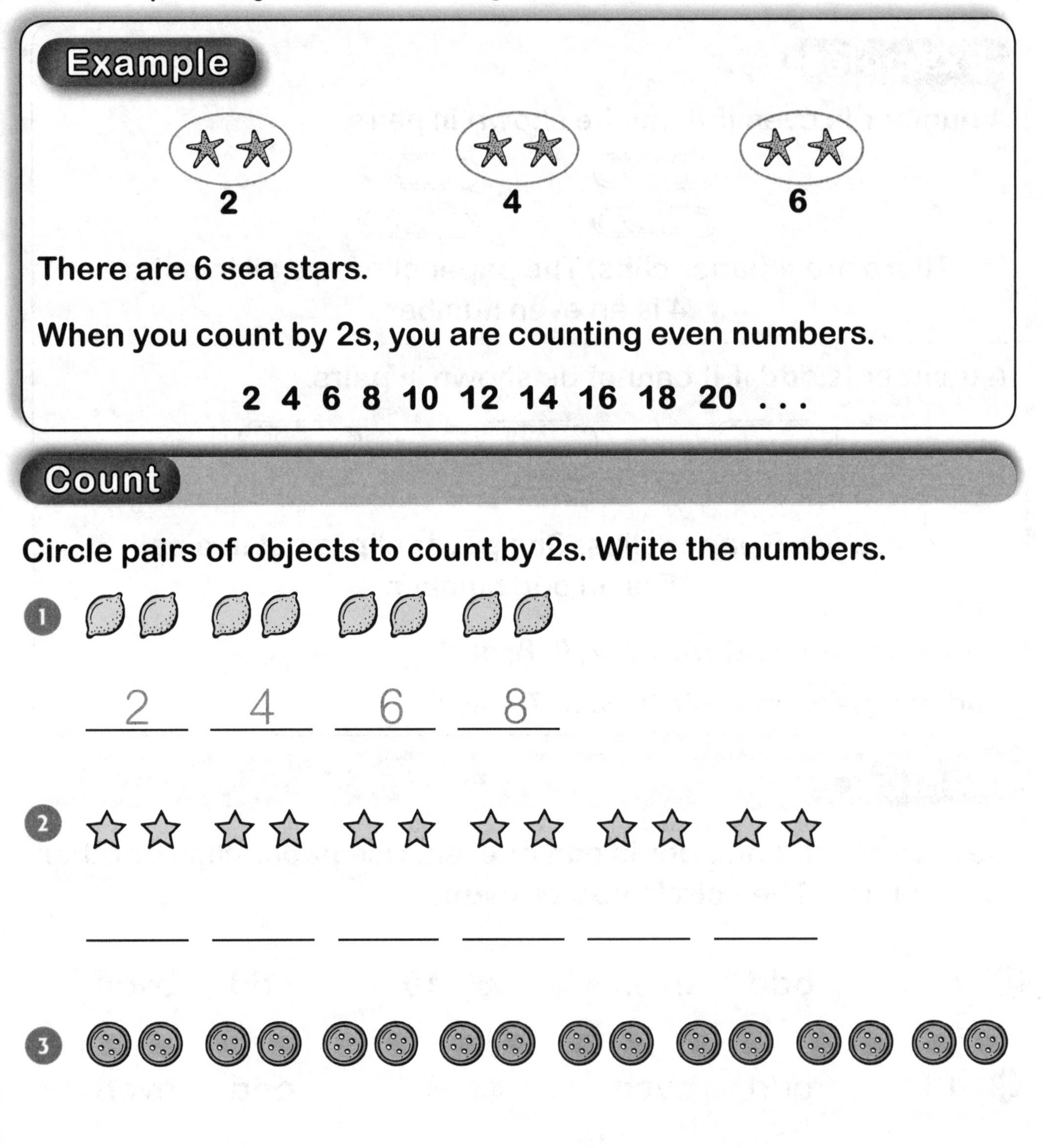

Example

2 4 6

There are 6 sea stars.

When you count by 2s, you are counting even numbers.

2 4 6 8 10 12 14 16 18 20 . . .

Count

Circle pairs of objects to count by 2s. Write the numbers.

1. 2 4 6 8

2. ______ ______ ______ ______ ______ ______

3. ______ ______ ______ ______ ______ ______ ______ ______

Name ____________________

Even and Odd Sums

A sum can be even or odd.

Examples

A sum is an even number when both addends are even numbers.

2 + 4 = 6
2 and 4 are both even numbers.
The sum is an even number.

A sum is an even number when both addends are odd numbers.

3 + 9 = 12
3 and 9 are both odd numbers.
The sum is an even number.

A sum is an odd number when one of the addends is even and the other is odd.

3 + 4 = 7
3 is an odd number. 4 is an even number.
The sum is an odd number.

Identify

Write the sum. Tell if the sum is even or odd.

1. 5 + 5 = 10 even
2. 12 + 7 = ______ ______
3. 6 + 8 = ______ ______
4. 5 + 9 = ______ ______

Lesson 4

Name ____________________

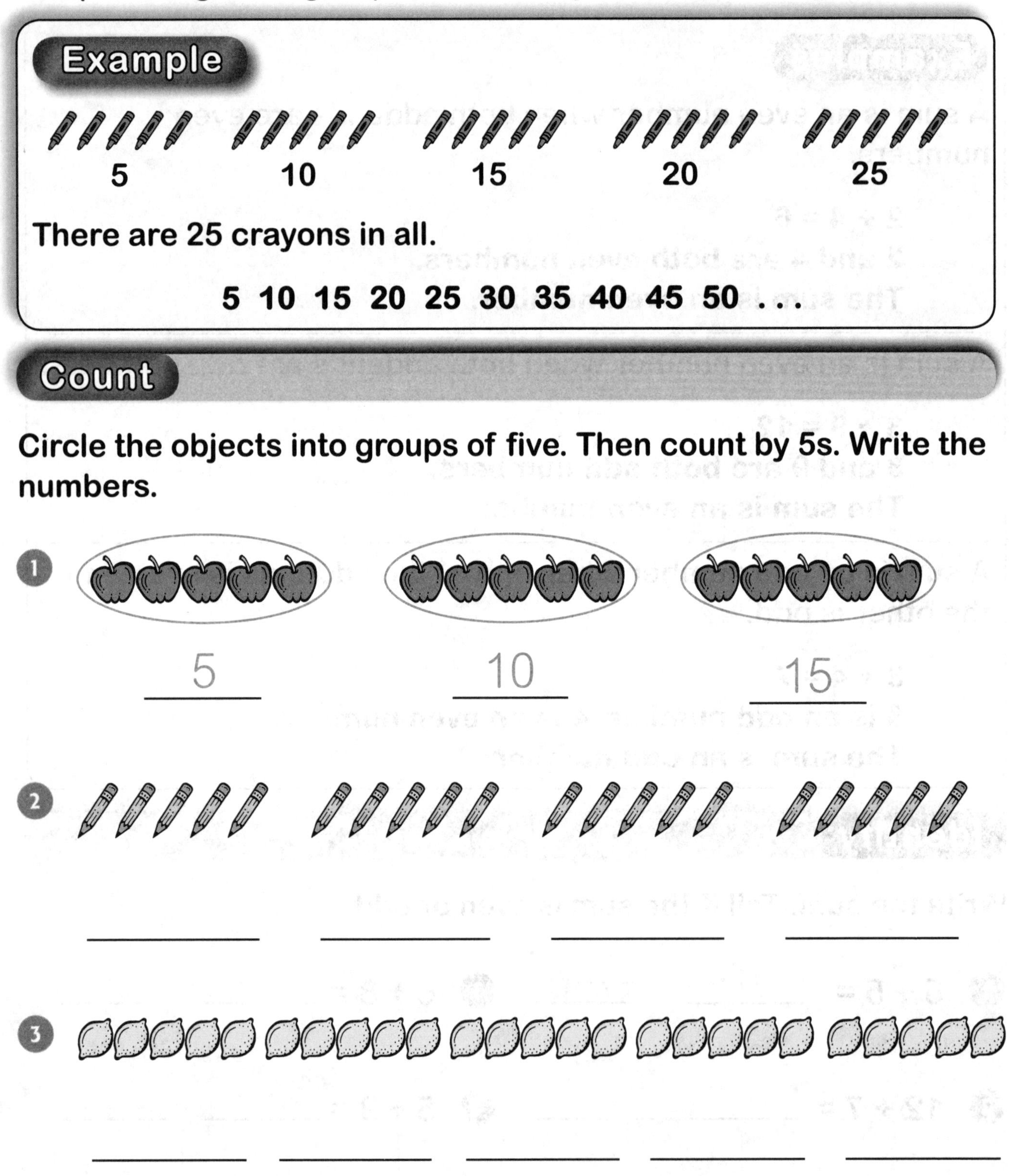

Counting by 5s

You put things into groups of 5. Then you can count by 5s.

Example

5 10 15 20 25

There are 25 crayons in all.

5 10 15 20 25 30 35 40 45 50 . . .

Count

Circle the objects into groups of five. Then count by 5s. Write the numbers.

1. 5 10 15

2. ______ ______ ______ ______

3. ______ ______ ______ ______ ______

Arrays

Arrays show equal groups. They can help you add.

Examples

Arrays have rows with the same number of objects.

There are 3 rows. There are 4 pennies in each row.

An addition sentence can tell what an array shows.
There are 3 rows of 4 pennies.

4 pennies + 4 pennies + 4 pennies = 12 pennies in all

Add

Look at the array. Write the sum.

1. 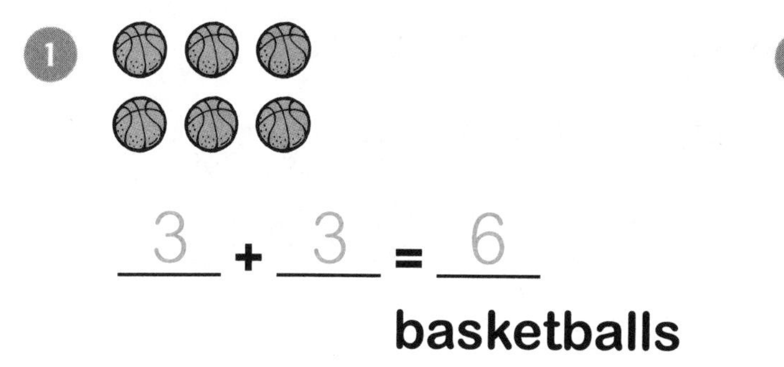

   3 + 3 = 6 basketballs

2. 5 + 5 + 5 = ____ fish

Name ____________________

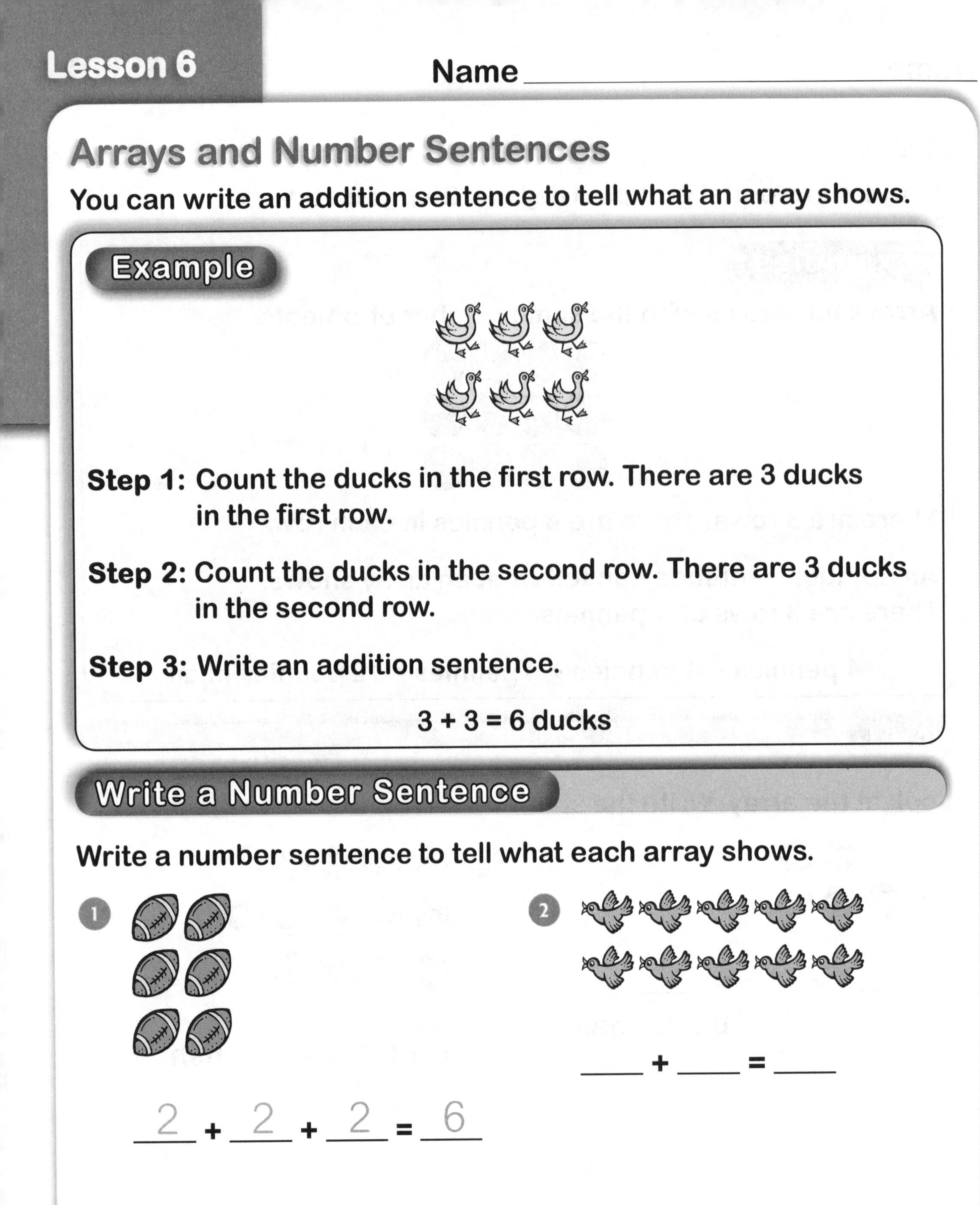

Arrays and Number Sentences

You can write an addition sentence to tell what an array shows.

Example

Step 1: Count the ducks in the first row. There are 3 ducks in the first row.

Step 2: Count the ducks in the second row. There are 3 ducks in the second row.

Step 3: Write an addition sentence.

3 + 3 = 6 ducks

Write a Number Sentence

Write a number sentence to tell what each array shows.

1. 2 + 2 + 2 = 6

2. ____ + ____ = ____

Name ______________________

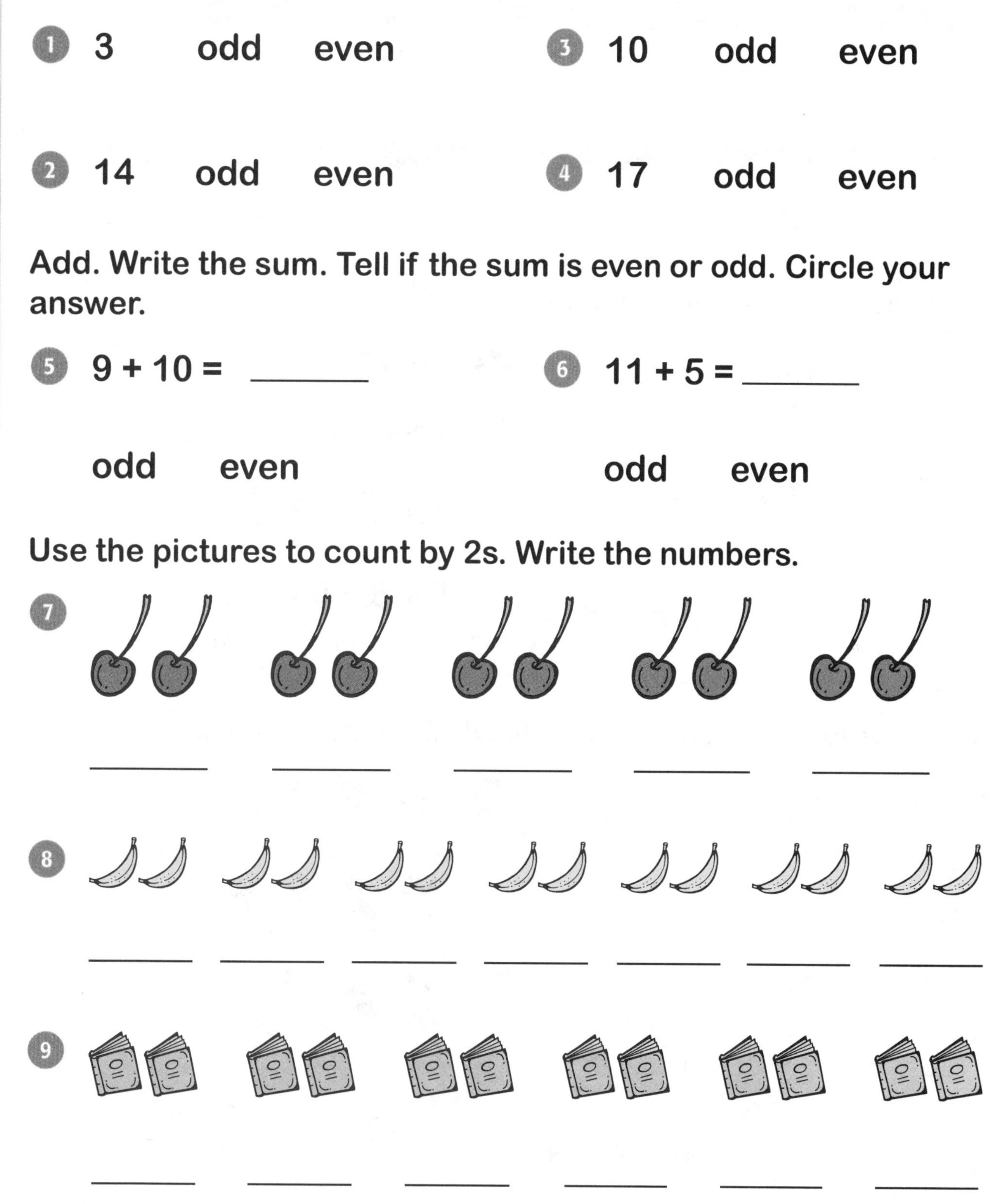

Is the number odd or even? Circle your answer. Use small objects to help.

1. 3 odd even
2. 14 odd even
3. 10 odd even
4. 17 odd even

Add. Write the sum. Tell if the sum is even or odd. Circle your answer.

5. 9 + 10 = ______

   odd even

6. 11 + 5 = ______

   odd even

Use the pictures to count by 2s. Write the numbers.

7. ______ ______ ______ ______ ______

8. ______ ______ ______ ______ ______ ______ ______

9. ______ ______ ______ ______ ______ ______

Name ____________________

Use the pictures to count by 5s. Write the numbers.

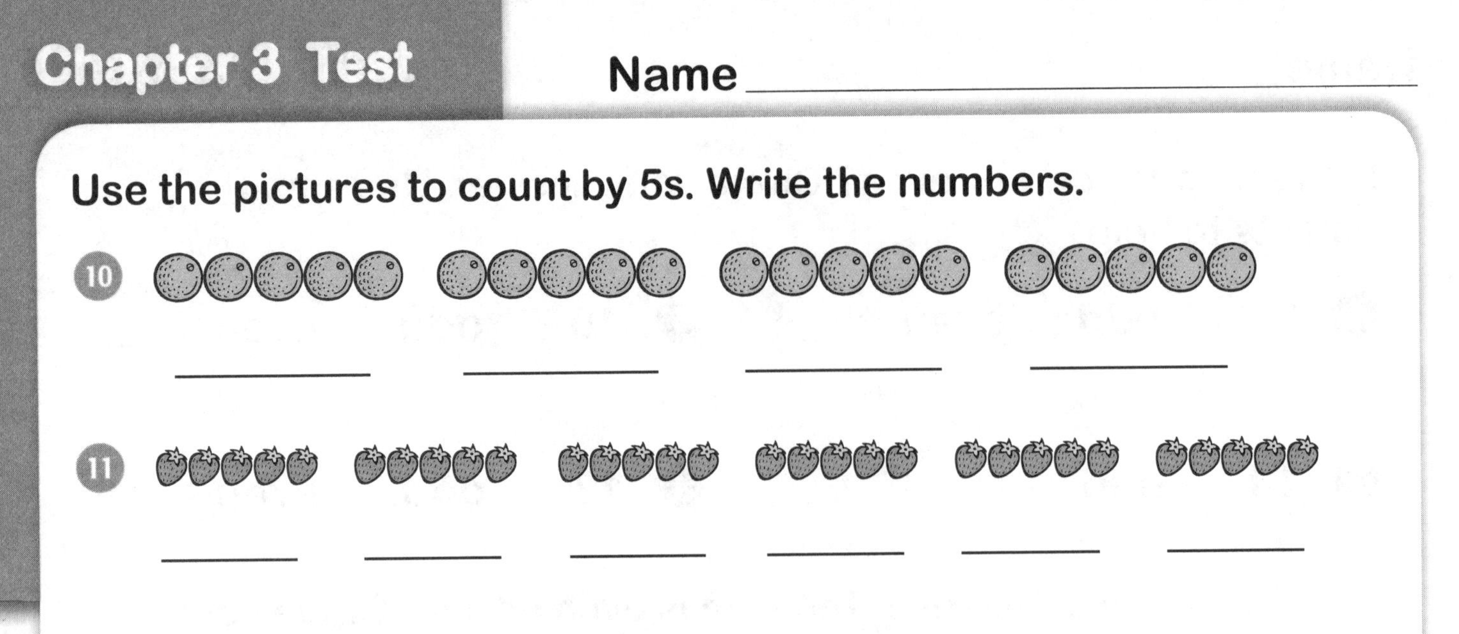

10 ________ ________ ________ ________

11 ______ ______ ______ ______ ______ ______

Look at the array. Write the sum.

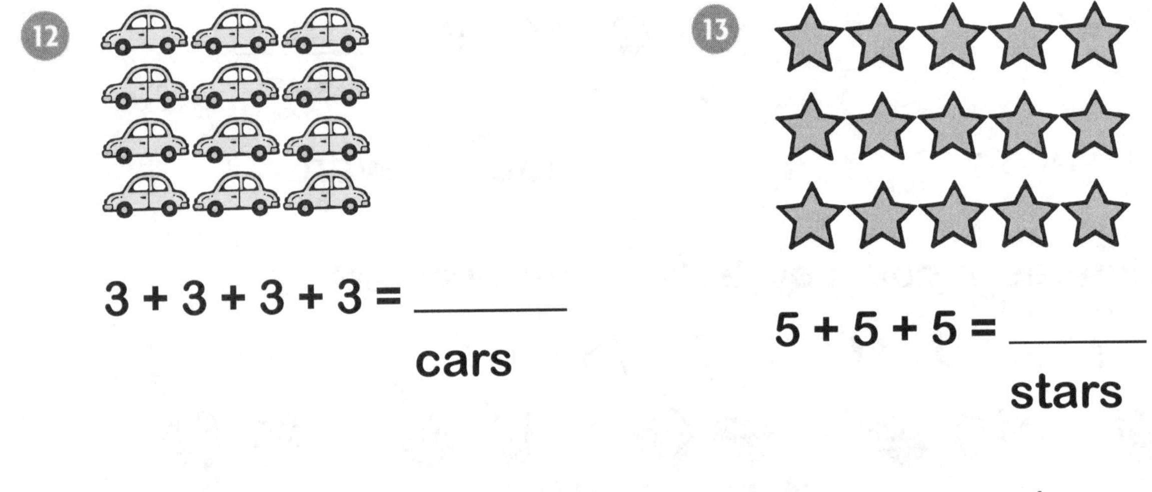

12 $3 + 3 + 3 + 3 =$ ______ cars

13 $5 + 5 + 5 =$ ______ stars

Write a number sentence that tells what the array shows.

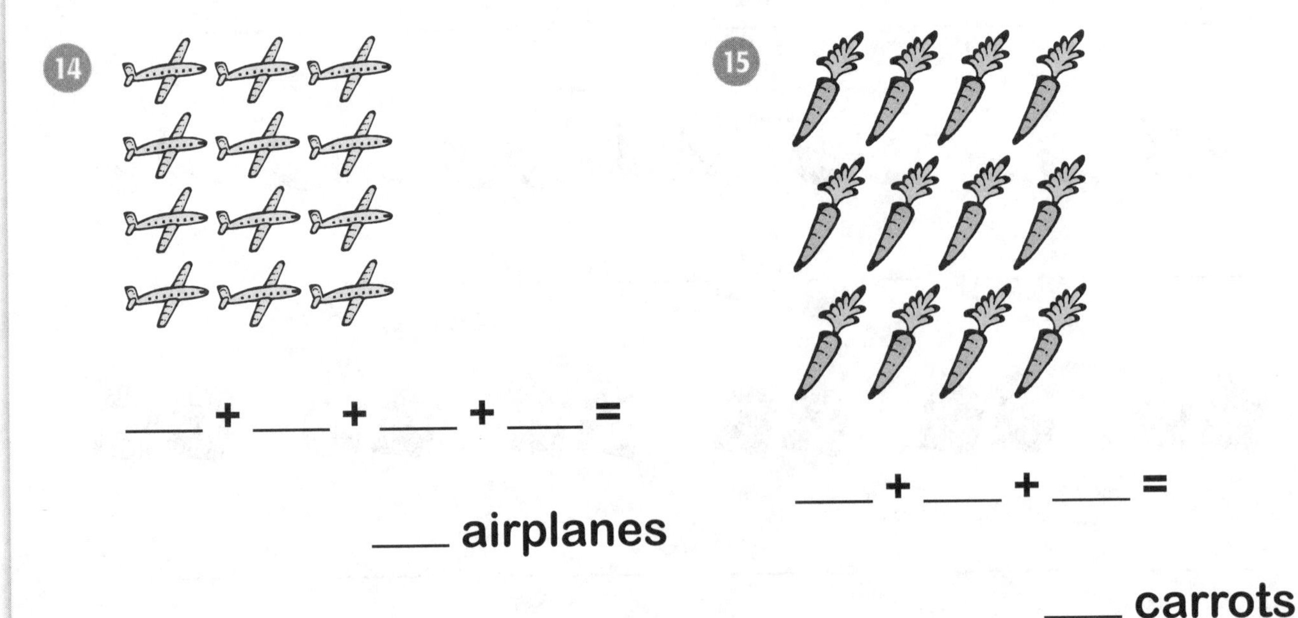

14 ___ + ___ + ___ + ___ =

___ airplanes

15 ___ + ___ + ___ =

___ carrots

Name ______________________

Counting Hundreds

Three-digit numbers are made of hundreds, tens, and ones.

Example

You can use number cubes to show numbers.

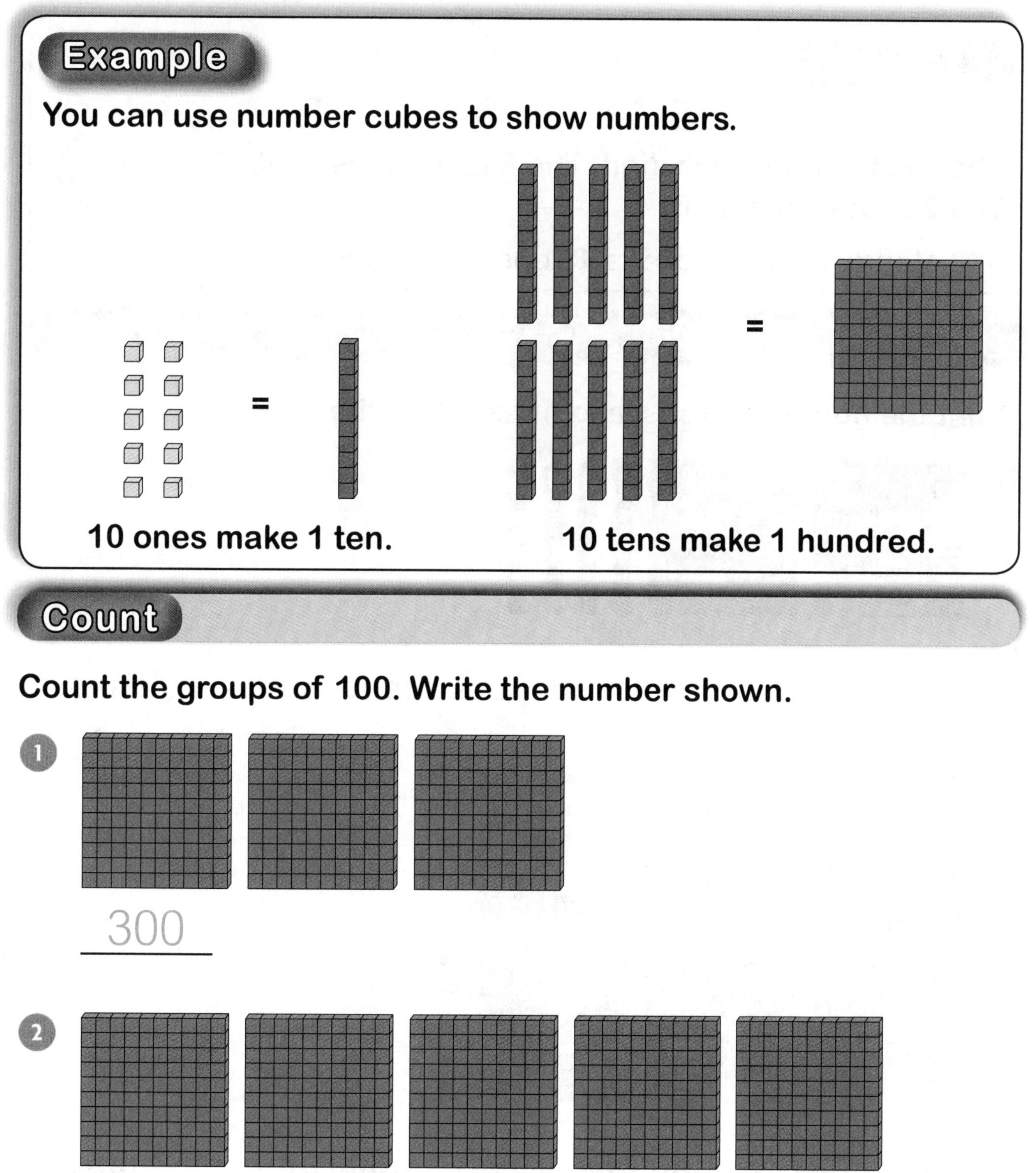

10 ones make 1 ten.

10 tens make 1 hundred.

Count

Count the groups of 100. Write the number shown.

1. 300

2. ________

Name ____________________

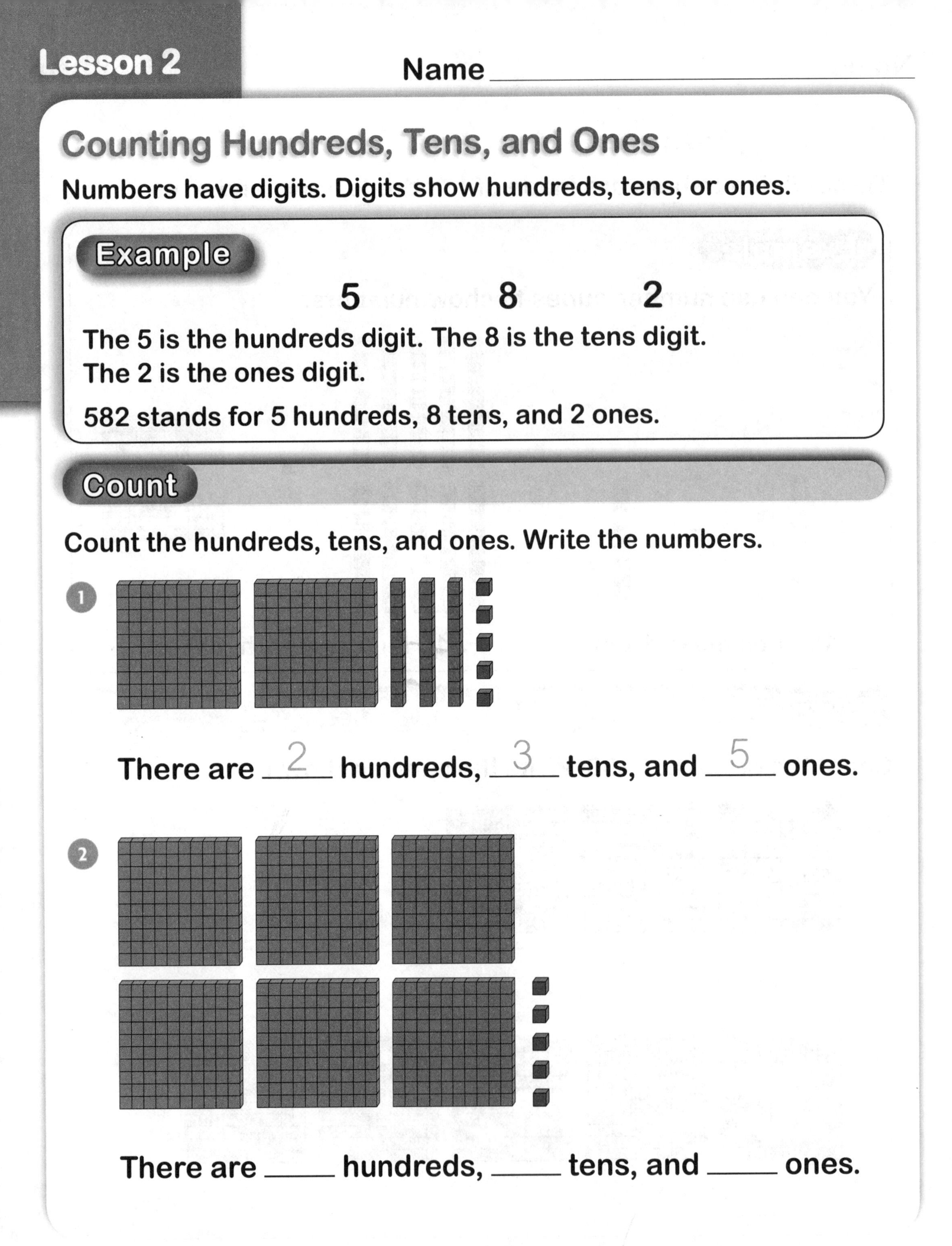

Counting Hundreds, Tens, and Ones

Numbers have digits. Digits show hundreds, tens, or ones.

Example

5 8 2

The 5 is the hundreds digit. The 8 is the tens digit.
The 2 is the ones digit.

582 stands for 5 hundreds, 8 tens, and 2 ones.

Count

Count the hundreds, tens, and ones. Write the numbers.

1. There are __2__ hundreds, __3__ tens, and __5__ ones.

2. There are _____ hundreds, _____ tens, and _____ ones.

Name ______________________________

Skip Counting by 5s

You can skip count by 5s.

Examples

When you count by 5s, you count in a pattern.

0 5 10 15 20 25 30 ...

The number increases by 5 each time.

Start at 265. Skip count by 5.

265 270 275 280 285 290 295 ...

Count

Count by 5s. Write the numbers.

1. 185 190 195 200 205 210 215

2. 635 640 ______ ______ ______ ______ ______

3. 520 525 ______ ______ ______ ______ ______

4. 785 790 ______ ______ ______ ______ ______

Lesson 4

Name ______________________

Skip Counting by 10s

You can count by 10s.

Examples

When you count by 10s, you count in a pattern.

0 10 20 30 40 50 60 ...

The number increases by 10 each time.

260 270 280 290 300 310 320 330 340 ...

Count

Count by 10s. Write the numbers.

1. 200 210 220 230 240 250 260

2. 340 350 ______ ______ ______ ______ ______

3. 820 830 ______ ______ ______ ______ ______

4. 910 920 ______ ______ ______ ______ ______

5. 740 750 ______ ______ ______ ______ ______

Name ______________________

Skip Counting by 100s

You can count by 100s.

Examples

When you count by 100s, you count in a pattern.

0 100 200 300 400 500 600 ...

The number increases by 100 each time.

175 275 375 475 575 675 775 875 ...

Count

Count by 100s. Write the numbers.

1. 300 400 500 600 700 800 900

2. 185 285 ______ ______ ______ ______ ______

3. 350 450 ______ ______ ______ ______ ______

4. 220 320 ______ ______ ______ ______ ______

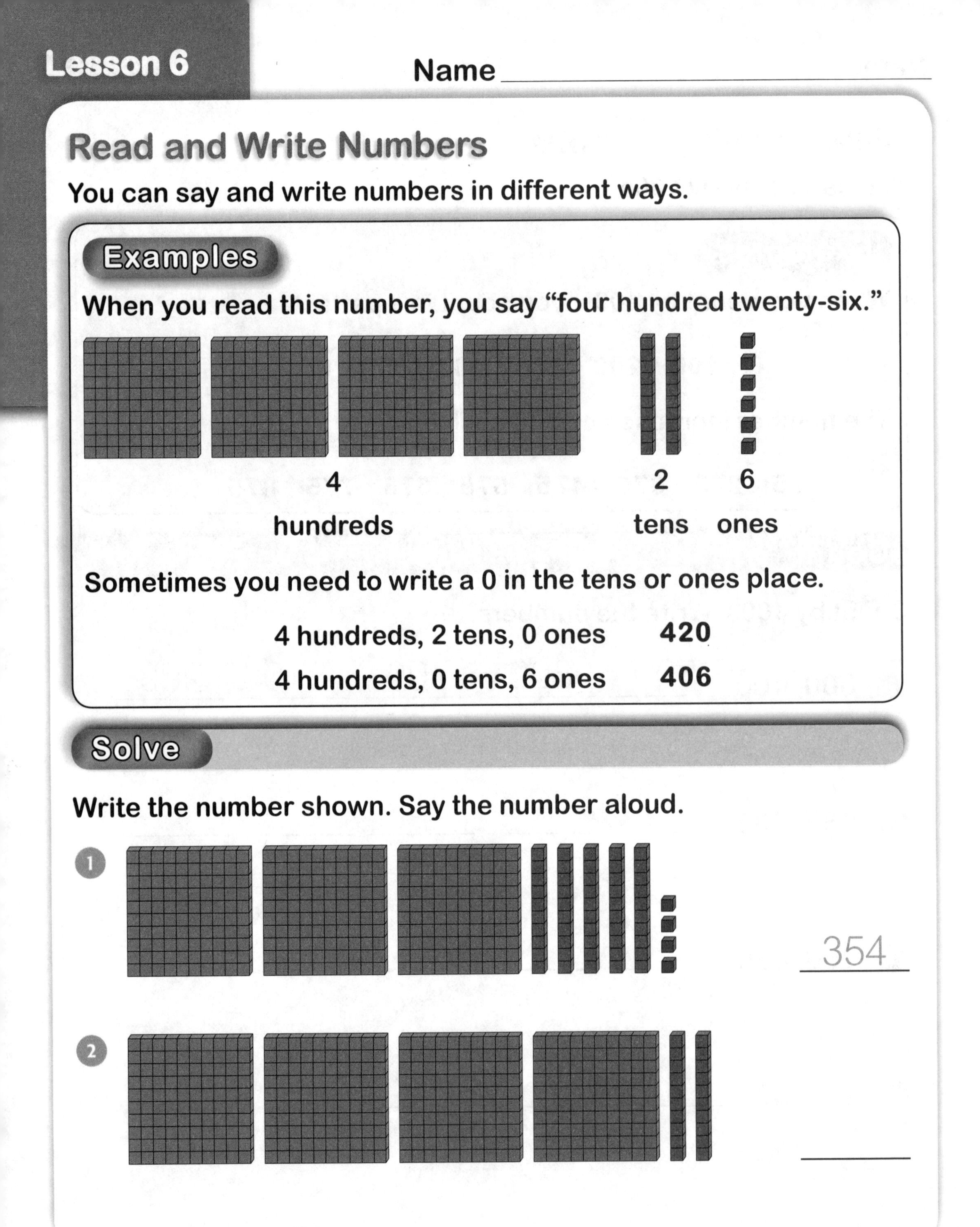

Name ______________________

Read and Write Numbers

You can say and write numbers in different ways.

Examples

When you read this number, you say "four hundred twenty-six."

4	2	6
hundreds	tens	ones

Sometimes you need to write a 0 in the tens or ones place.

4 hundreds, 2 tens, 0 ones **420**

4 hundreds, 0 tens, 6 ones **406**

Solve

Write the number shown. Say the number aloud.

1. 354

2. ______

Name ______________________________

Read and Write Numbers Using Number Names

You can write words to show numbers.

Examples

This is how you read and write numbers for the tens and ones places.

10	ten	15	fifteen	20	twenty	60	sixty
11	eleven	16	sixteen	30	thirty	70	seventy
12	twelve	17	seventeen	40	forty	80	eighty
13	thirteen	18	eighteen	50	fifty	90	ninety
14	fourteen	19	nineteen				

Put a hyphen between the words for numbers in the tens and ones places.

129 is **one hundred twenty-nine.**

467 is **four hundred sixty-seven.**

Write

Read the number. Write the number in words.

1. 325 three hundred twenty-five
2. 842 ______________________________
3. 699 ______________________________
4. 730 ______________________________
5. 105 ______________________________
6. 517 ______________________________

Lesson 8

Name ____________________

Read and Write Numbers in Expanded Form

You can write numbers in expanded form. Expanded form shows the value of each digit.

Example

This picture shows hundreds, tens, and ones. There are 3 hundreds, 2 tens, and 8 ones.

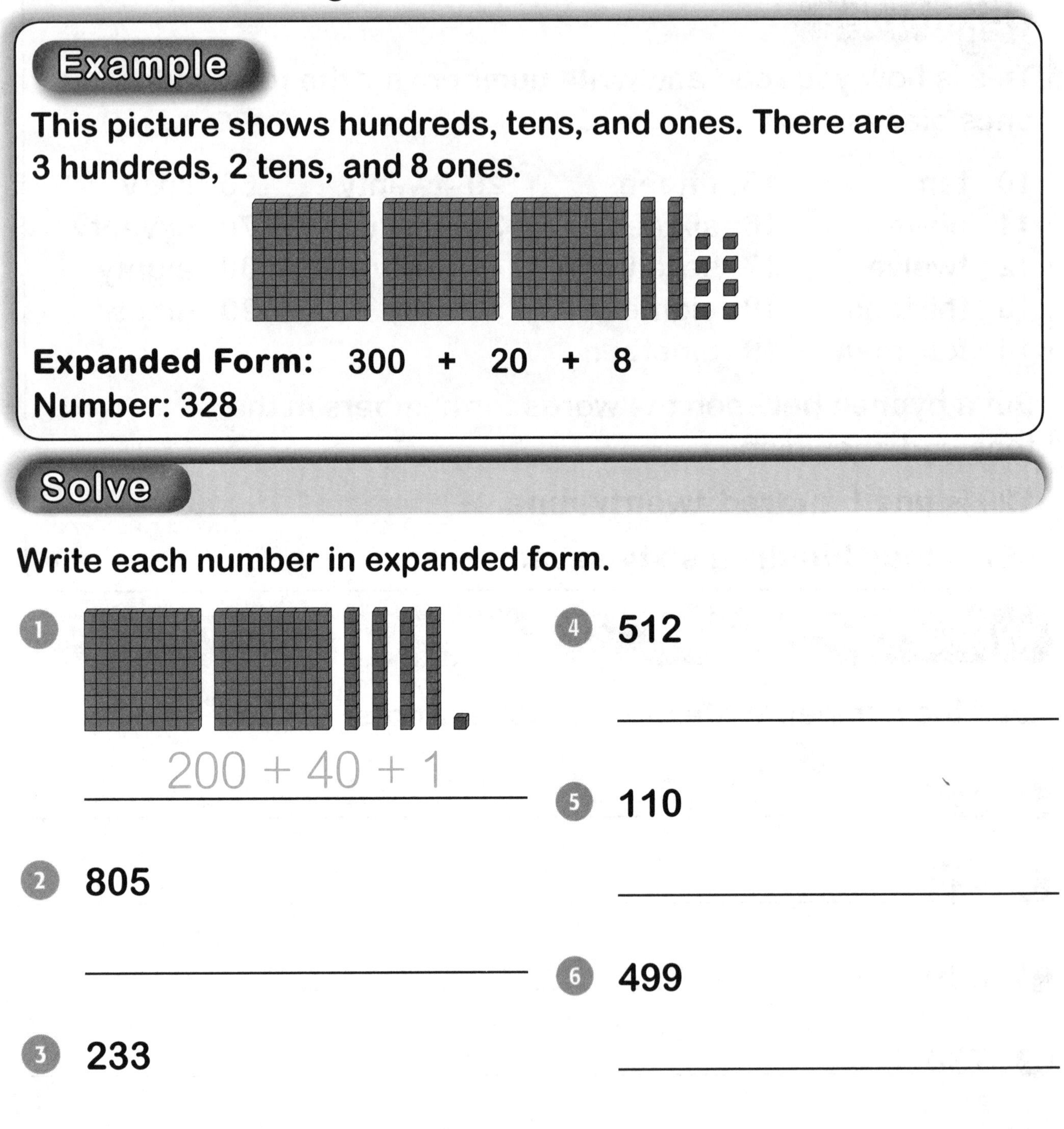

Expanded Form: 300 + 20 + 8

Number: 328

Solve

Write each number in expanded form.

1. 200 + 40 + 1

2. 805 ____________________

3. 233 ____________________

4. 512 ____________________

5. 110 ____________________

6. 499 ____________________

Name ______________________________

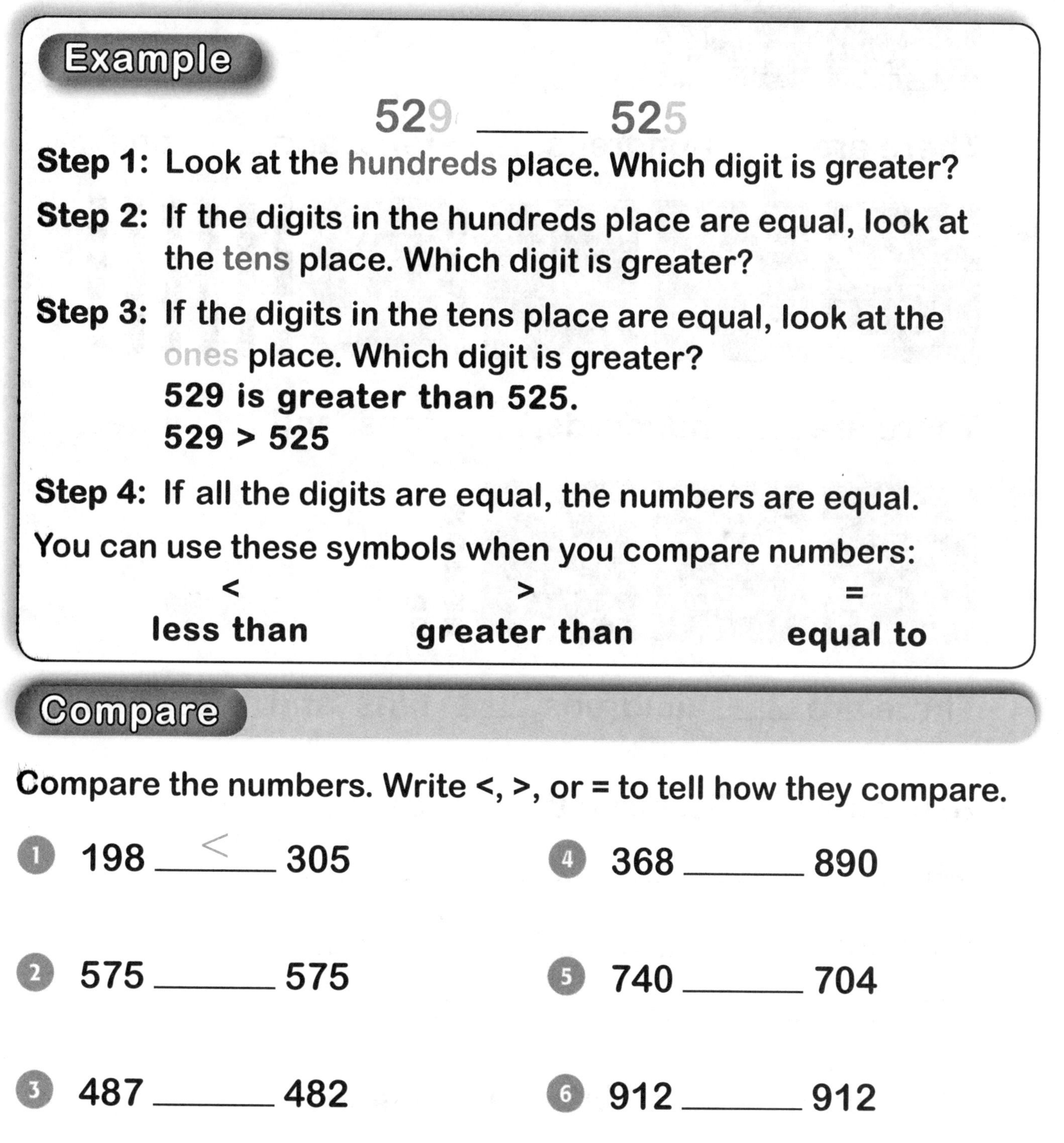

Comparing Numbers Using Symbols

When you compare numbers, you tell if a number is less than, greater than, or equal to another number.

Example

529 _____ 525

Step 1: Look at the hundreds place. Which digit is greater?

Step 2: If the digits in the hundreds place are equal, look at the tens place. Which digit is greater?

Step 3: If the digits in the tens place are equal, look at the ones place. Which digit is greater?
529 is greater than 525.
529 > 525

Step 4: If all the digits are equal, the numbers are equal.

You can use these symbols when you compare numbers:

<	>	=
less than	**greater than**	**equal to**

Compare

Compare the numbers. Write <, >, or = to tell how they compare.

1. 198 ___<___ 305
2. 575 ______ 575
3. 487 ______ 482
4. 368 ______ 890
5. 740 ______ 704
6. 912 ______ 912

Name ______________________

Count the hundreds, tens, and ones. Write the numbers.

1. 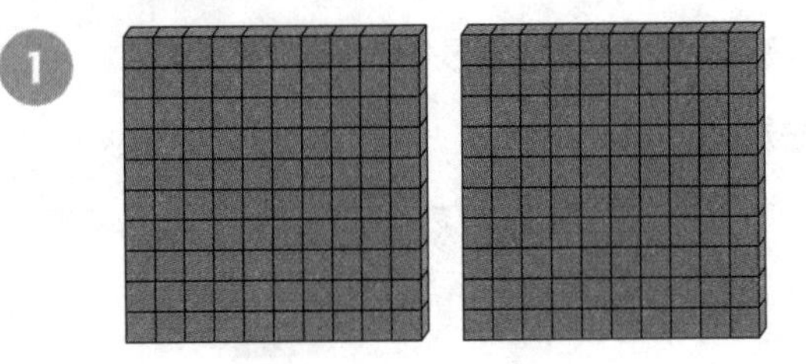

There are _____ hundreds, _____ tens, and _____ ones.

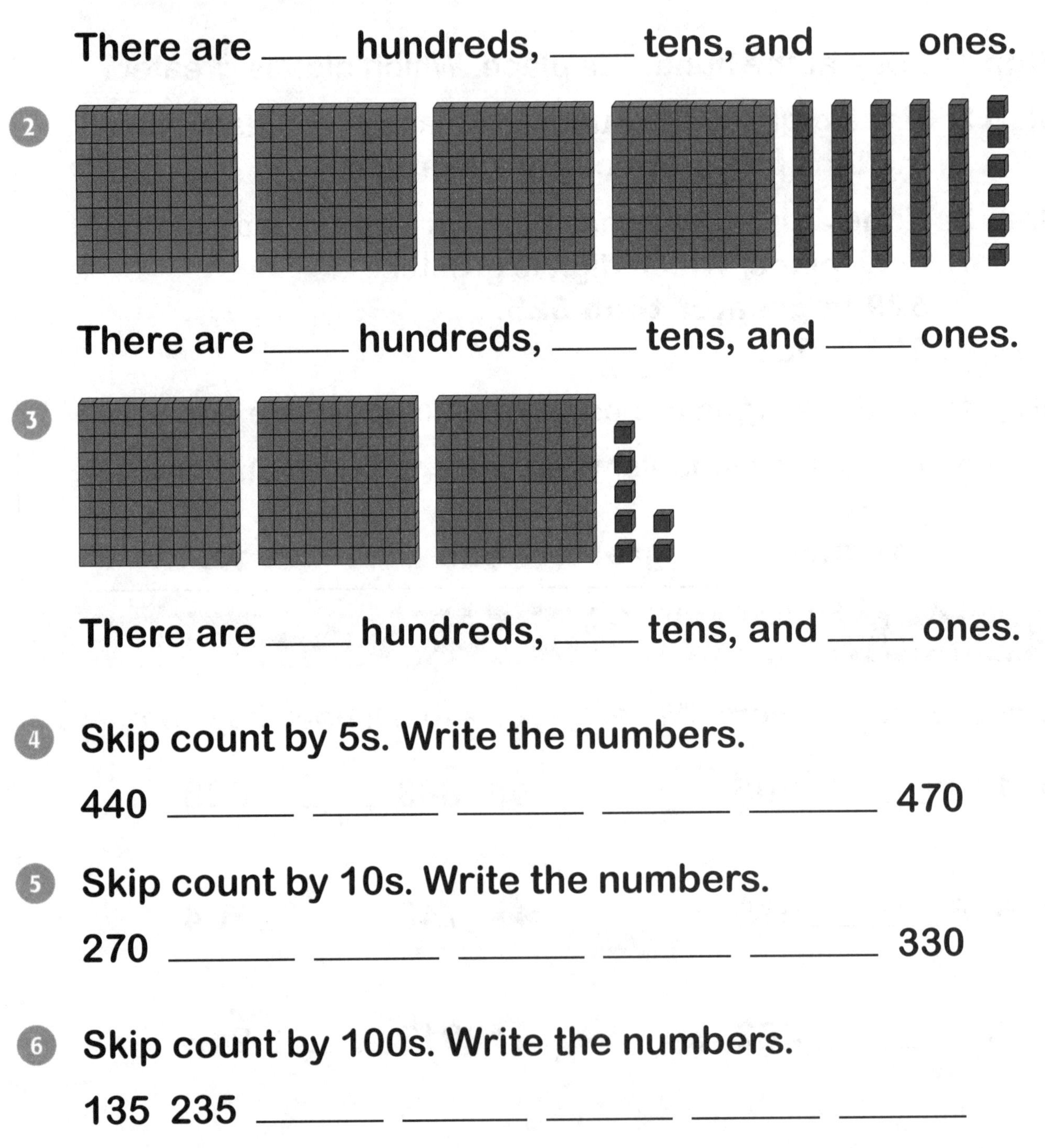

2. There are _____ hundreds, _____ tens, and _____ ones.

3. There are _____ hundreds, _____ tens, and _____ ones.

4. Skip count by 5s. Write the numbers.

440 _______ _______ _______ _______ _______ 470

5. Skip count by 10s. Write the numbers.

270 _______ _______ _______ _______ _______ 330

6. Skip count by 100s. Write the numbers.

135 235 _______ _______ _______ _______ _______

Name ______________________

7. Write the number shown. Say the number aloud.

____

8. Write the number in words.

____

9. Write the number in expanded form.

____

Compare each set of numbers. Write <, >, or =.

10. 285 ______ 122

11. 68 ______ 77

12. 748 ______ 748

13. 446 ______ 546

14. 345 ______ 345

15. 192 ______ 219

16. 462 ______ 437

17. 552 ______ 498

18. 267 ______ 276

19. 978 ______ 977

Chapters 1-4 Review

Name ____________________

Add or subtract. Write the sum or difference.

1. $2 + 4 =$ ______
2. $11 + 3 =$ ______
3. $9 + 8 =$ ______
4. $4 + 2 + 7 =$ ______
5. $7 - 3 =$ ______
6. $10 - 10 =$ ______
7. $18 - 4 =$ ______
8. $14 - 9 =$ ______

Solve. Write the number sentence. Then write the answer.

9. Lucy catches 4 .
Her mom catches 9 .
How many do they catch in all?

______ + ______ = ______

10. Ken has 8 .
He gives 6 to his brother.
Then his dad gives Ken 7 more .
How many does Ken have now?

______ – ______ = ______

______ + ______ = ______

11. There are 15 on a plate.
Annie takes 6 .
How many are left?

______ – ______ = ______

12. A store has 6 .
It also has 10 .
The store sells 8 kites.
How many kites are left?

______ + ______ = ______ kites

______ – ______ = ______ kites

Name ____________________

Draw a picture and write a number sentence to solve.
Use another sheet of paper for your drawing.

13. Ms. Chan has 11 black buttons. She has 6 brown buttons. How many buttons does she have in all?

______ ◯ ______ = ?

______ buttons

14. Jill has 14 toy cars. She gives 8 of them to her sister. How many cars does Jill have now?

______ toy cars

Add or subtract. Write the sum or difference.

15. $34 + 15$

16. $22 + 64$

17. $38 + 27$

18. $56 + 18$

19. $58 - 23$

20. $46 - 25$

21. $63 - 17$

22. $81 - 28$

23. $96 - 29$

Name ______________________

Write the number sentence for each exercise. Then solve and write your answer.

24 There are 35 books on a shelf.
Ms. Jones takes 9 books.
How many books are left?

______ books

25 June has 67 blocks.
Kevin has 48 blocks.
How many more blocks does June have than Kevin?

______ blocks

26 A store has 48 bikes.
The store sells 20 bikes.
The store gets 35 more bikes.
How many bikes are in the store now?

______ bikes

27 Mr. Lewis has 83 nails.
He uses 56 of them.
He gets 30 more nails.
How many nails does he have now?

______ nails

28 Bob has 28 stamps.
Nick has 53 stamps.
They use 47 stamps.
How many stamps are left over?

______ stamps

29 A store has a basket with 44 apples in it.
Another basket has 19 apples.
The store sells 40 apples.
How many apples does the store still have?

______ apples

Name ______________________

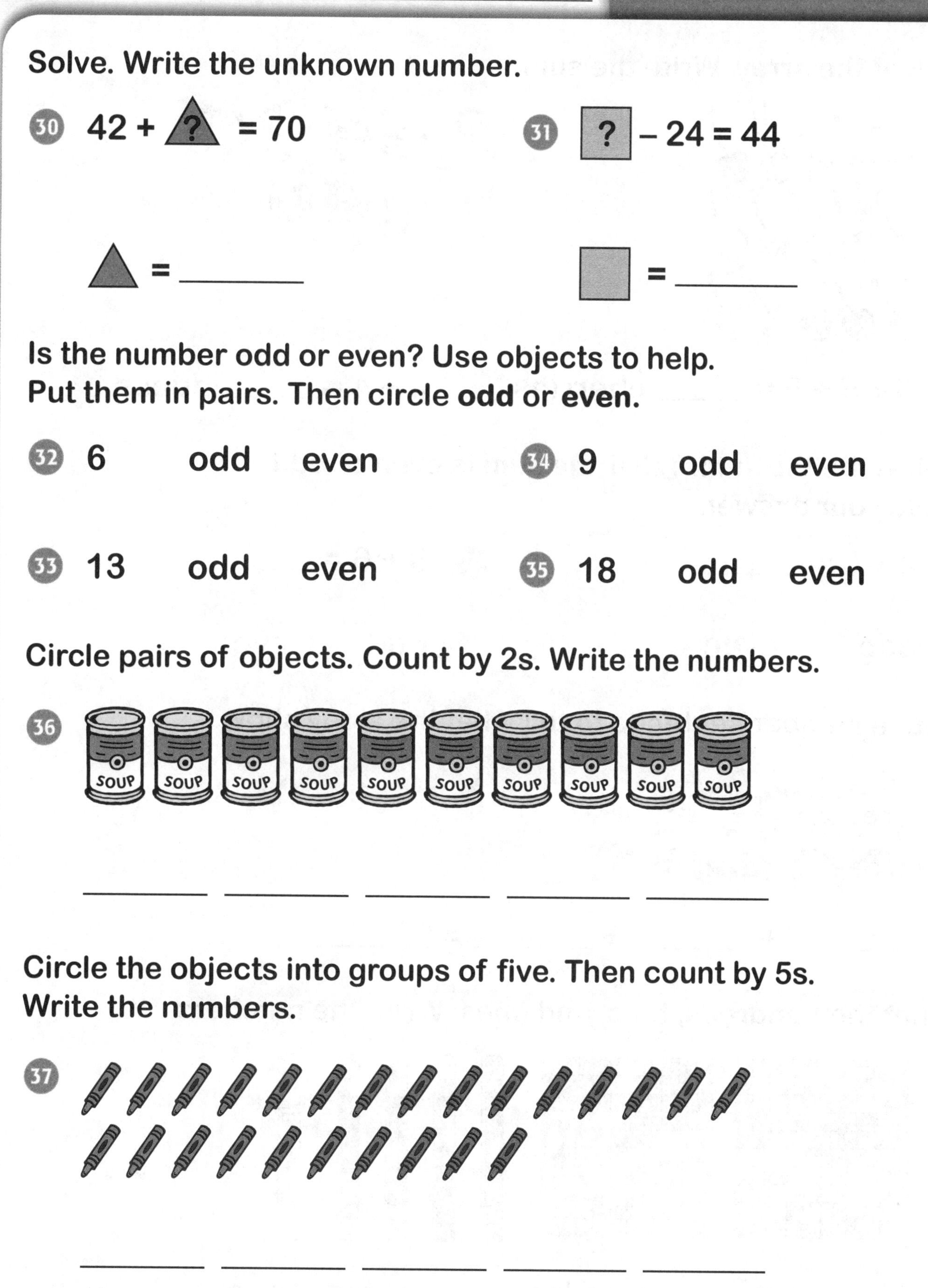

Solve. Write the unknown number.

30 $42 + \triangle = 70$

$\triangle$ = ______

31 $\square - 24 = 44$

$\square$ = ______

Is the number odd or even? Use objects to help.
Put them in pairs. Then circle **odd** or **even.**

32 6 odd even

33 13 odd even

34 9 odd even

35 18 odd even

Circle pairs of objects. Count by 2s. Write the numbers.

36

______ ______ ______ ______ ______

Circle the objects into groups of five. Then count by 5s.
Write the numbers.

37

______ ______ ______ ______ ______

Name ________________________________

Look at the array. Write the sum.

38

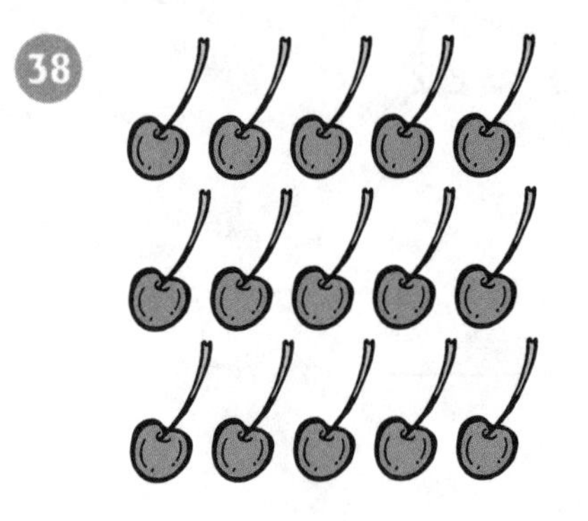

$5 + 5 + 5 =$ _____ cherries

39

$4 + 4 =$ _____ ducks

Add. Write the sum. Tell if the sum is even or odd. Circle your answer.

40 $4 + 5 =$ _______

odd even

41 $8 + 6 =$ _______

odd even

Write a number sentence to tell what the array shows.

42

_______ + _______ + _______ = _______

Count the hundreds, tens, and ones. Write the numbers.

43

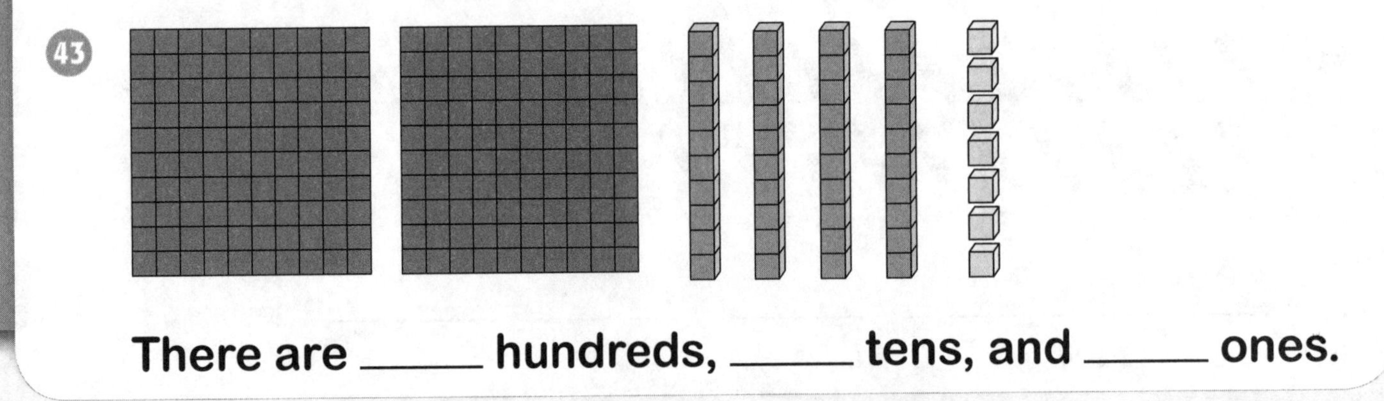

There are _____ hundreds, _____ tens, and _____ ones.

Name ___________________________

44 Skip count by 5s. Write the numbers.

245 250 _____ _____ _____ _____ _____

45 Skip count by 10s. Write the numbers.

300 310 _____ _____ _____ _____ _____

46 Skip count by 100s. Write the numbers.

300 400 _____ _____ _____ _____ _____

47 Write the number shown.

____

48 Write the number in words.

____

Compare each set of numbers. Write <, >, or = to tell how they compare.

49 460 _______ 535

51 819 _______ 819

50 327 _______ 314

52 396 _______ 392

Name ______________________

Adding in Any Order or Group

You can add numbers in any order. The sum is the same.

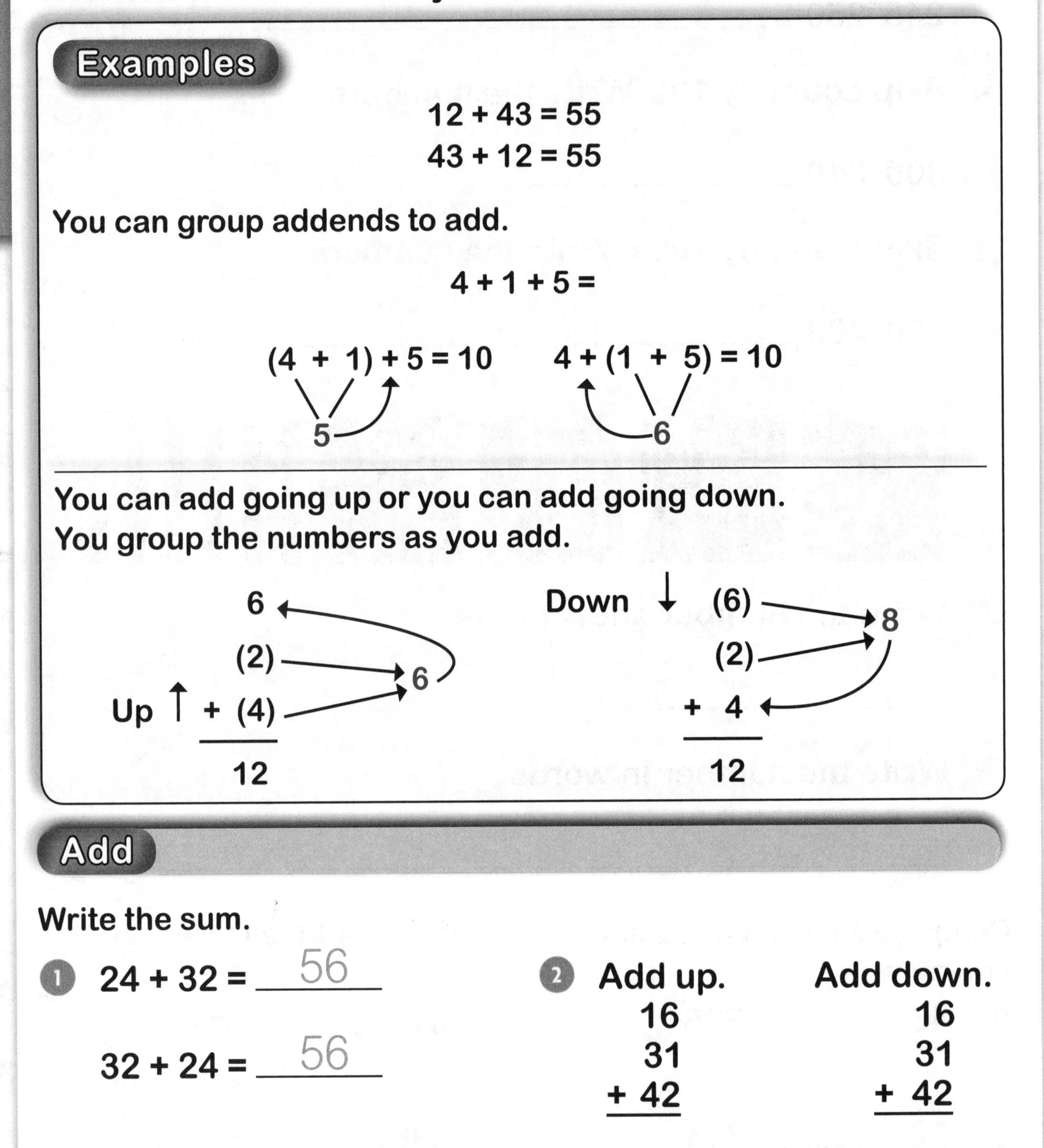

Examples

12 + 43 = 55
43 + 12 = 55

You can group addends to add.

4 + 1 + 5 =

(4 + 1) + 5 = 10   4 + (1 + 5) = 10

You can add going up or you can add going down.
You group the numbers as you add.

Add

Write the sum.

1. 24 + 32 = 56

   32 + 24 = 56

2. Add up.

   16
   31
   + 42

   Add down.

   16
   31
   + 42

Name ______________________________

Adding with Two, Three, or Four Addends

You can add four groups of numbers too.

Example

Brian scores 20 points in a basketball game.
Emma scores 18 points.
Nick scores 12 points.
Jose scores 16 points.
How many points did they score in all?

Number Sentence: 20 + 18 + 12 + 16 = 66

addends equals sum

You can order or group in different ways and then add. The sum is the same.

$$20 + (18 + 12) + 16 = 66$$

(18 + 12 = 30)

Add

Write the sum.

1. 8 + 4 + 7 + 2 = 21

   (8 + 4) + 7 + 2 = 21

2. 27 + 21 + 19 = ______

   (27 + 21) + 19 = ______

3. 30 + 11 + 55 = ______

   30 + (11 + 55) = ______

4. 53 + 19 = ______

   19 + 53 = ______

Name ____________________

Adding Three-Digit Numbers

You can add three-digit numbers.

Add 486 + 231

Step 1: Add the ones. Regroup if you need to.

Step 2: Add the tens. Regroup if you need to.

Step 3: Add the hundreds.

486 + 231 = 717

	Hundreds	Tens	Ones
	1		
	4	8	6
+	2	3	1
	7	1	7

The rules are the same with three addends.

Add

Write the sum. Regroup if you need to.

1. 611 + 208 = 819
2. 768 + 112
3. 331 + 225 + 281
4. 383 + 251 + 184
5. 101 + 89 + 403
6. 262 + 153 + 96
7. 700 + 146 + 111
8. 379 + 154 + 56

Name ______________________

Subtracting Three-Digit Numbers

You can subtract three-digit numbers.

Example

423 – 217

Step 1: Regroup if you need to. Subtract the ones.

Step 2: Regroup if you need to. Subtract the tens.

Step 3: Subtract the hundreds.

	Hundreds	Tens	Ones
	☐	1	13
	4	~~2~~	~~3~~
–	2	1	7
	2	0	6

423 – 217 = 206

Subtract

Write the difference. Regroup if you need to.

1. 819 – 611 = 208
2. 780 – 331 = ____
3. 880 – 112 = ____
4. 634 – 251 = ____
5. 458 – 399 = ____
6. 562 – 444 = ____
7. 950 – 95 = ____
8. 375 – 275 = ____

Name ____________________

Use Subtraction to Find an Unknown Addend

You can subtract to find a missing addend.

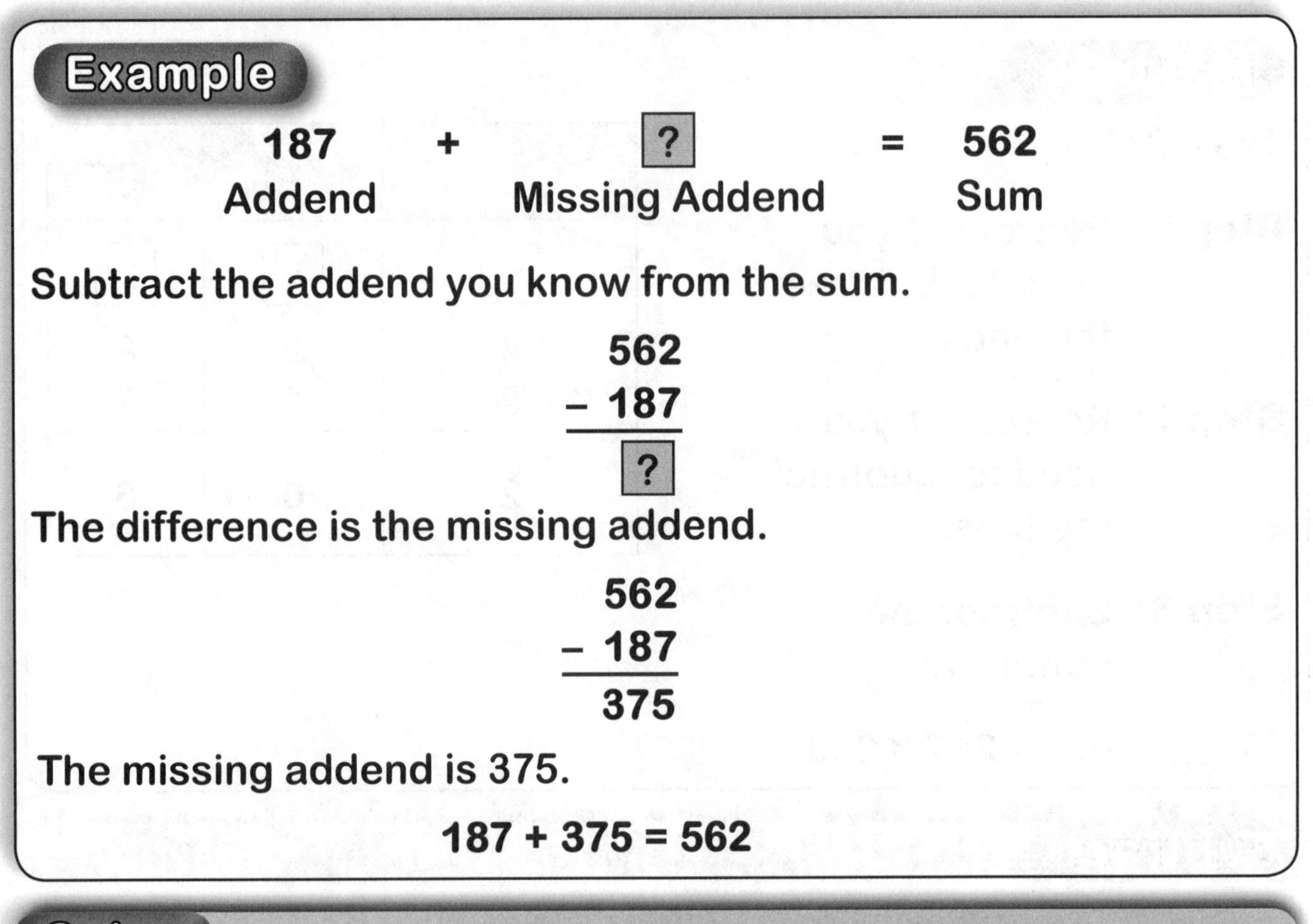

Example

187 + ? = 562

Addend, Missing Addend, Sum

Subtract the addend you know from the sum.

$$\begin{array}{r} 562 \\ -\ 187 \\ \hline ? \end{array}$$

The difference is the missing addend.

$$\begin{array}{r} 562 \\ -\ 187 \\ \hline 375 \end{array}$$

The missing addend is 375.

187 + 375 = 562

Solve

Subtract to find the missing addend.

1. 194 + ? = 422

$$\begin{array}{r} 422 \\ -\ 194 \\ \hline 228 \end{array}$$

194 + ________ = 422

2. 337 + ? = 612

$$\begin{array}{r} 612 \\ -\ 337 \\ \hline \end{array}$$

337 + ________ = 612

Name ______________________

Adding Three-Digit Numbers in Any Order or Group

You can add three-digit numbers in any order. The sum is the same.

Examples

```
  492        158
+ 158      + 492
  650        650
```

You can add going up or you can add going down.
You group the numbers as you add.

```
        1                          1
      129 ←─┐          Down ↓    12⑨ ──→ 11
      21② ──→ 7                  21② ──→ ┘
Up ↑ + 23⑤ ──→                 + 235 ←─┘
      576                        576
```

Add

Write the sum.

1.

```
  117        132
+ 132      + 117
  249
```

2.

```
  289        537
+ 537      + 289
```

3. Add up. Add down.

```
  225        142
  142        537
+ 537      + 225
```

4. Add up. Add down.

```
  104        327
  151        151
+ 327      + 104
```

Name ______________________________

Use Mental Math to Add 10s or 100s to a Number

You can easily add tens and hundreds to other numbers to find sums.

Examples

What is the sum of 228 and 30?

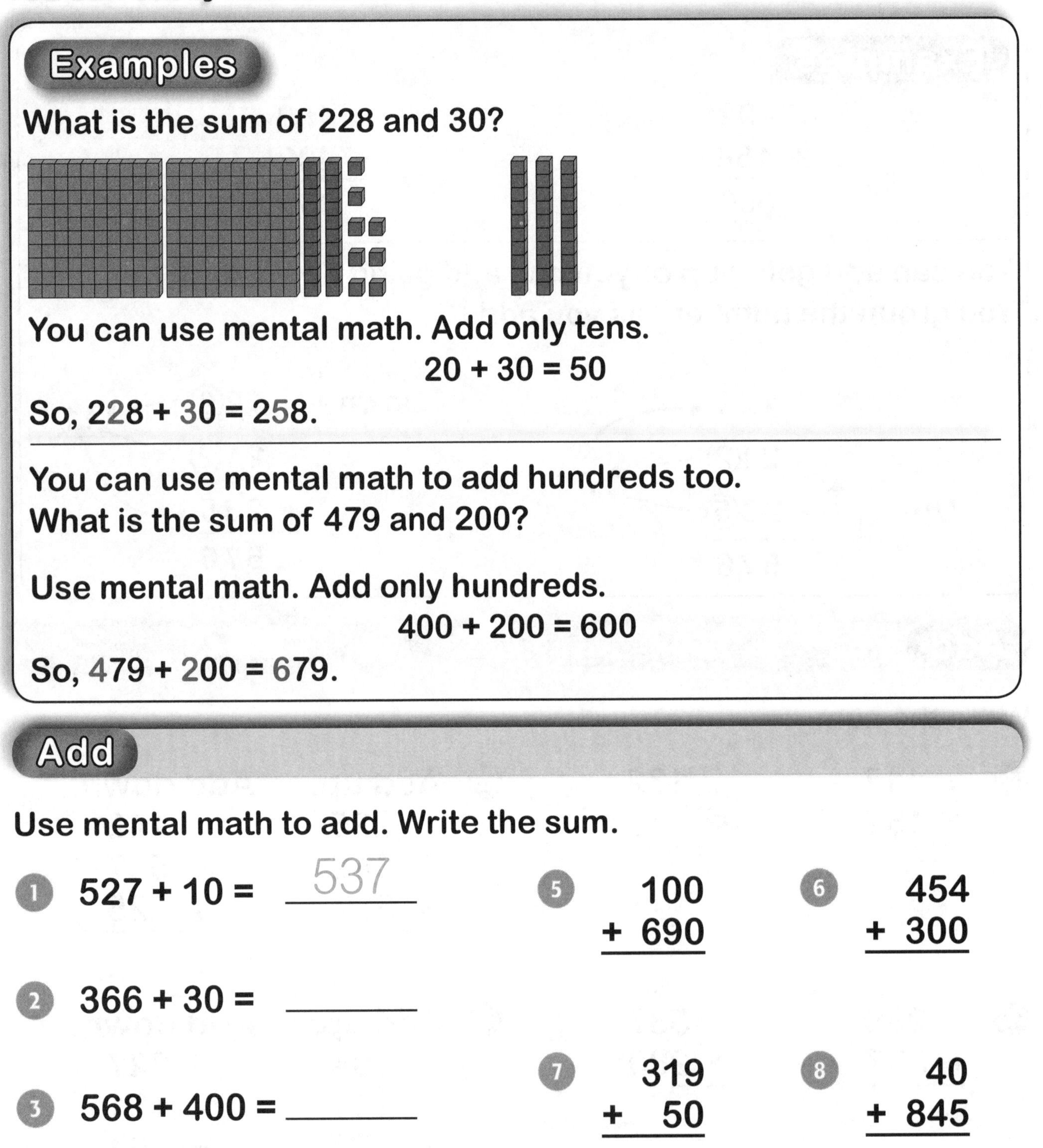

You can use mental math. Add only tens.

$$20 + 30 = 50$$

So, 228 + 30 = 258.

You can use mental math to add hundreds too.
What is the sum of 479 and 200?

Use mental math. Add only hundreds.

$$400 + 200 = 600$$

So, 479 + 200 = 679.

Add

Use mental math to add. Write the sum.

1. 527 + 10 = 537
2. 366 + 30 = ______
3. 568 + 400 = ______
4. 263 + 200 = ______
5. $\begin{array}{r} 100 \\ +\ 690 \\ \hline \end{array}$
6. $\begin{array}{r} 454 \\ +\ 300 \\ \hline \end{array}$
7. $\begin{array}{r} 319 \\ +\ \ 50 \\ \hline \end{array}$
8. $\begin{array}{r} 40 \\ +\ 845 \\ \hline \end{array}$

Name ______________________

Use Mental Math to Subtract 10s or 100s from a Number

You can easily subtract tens and hundreds to find differences.

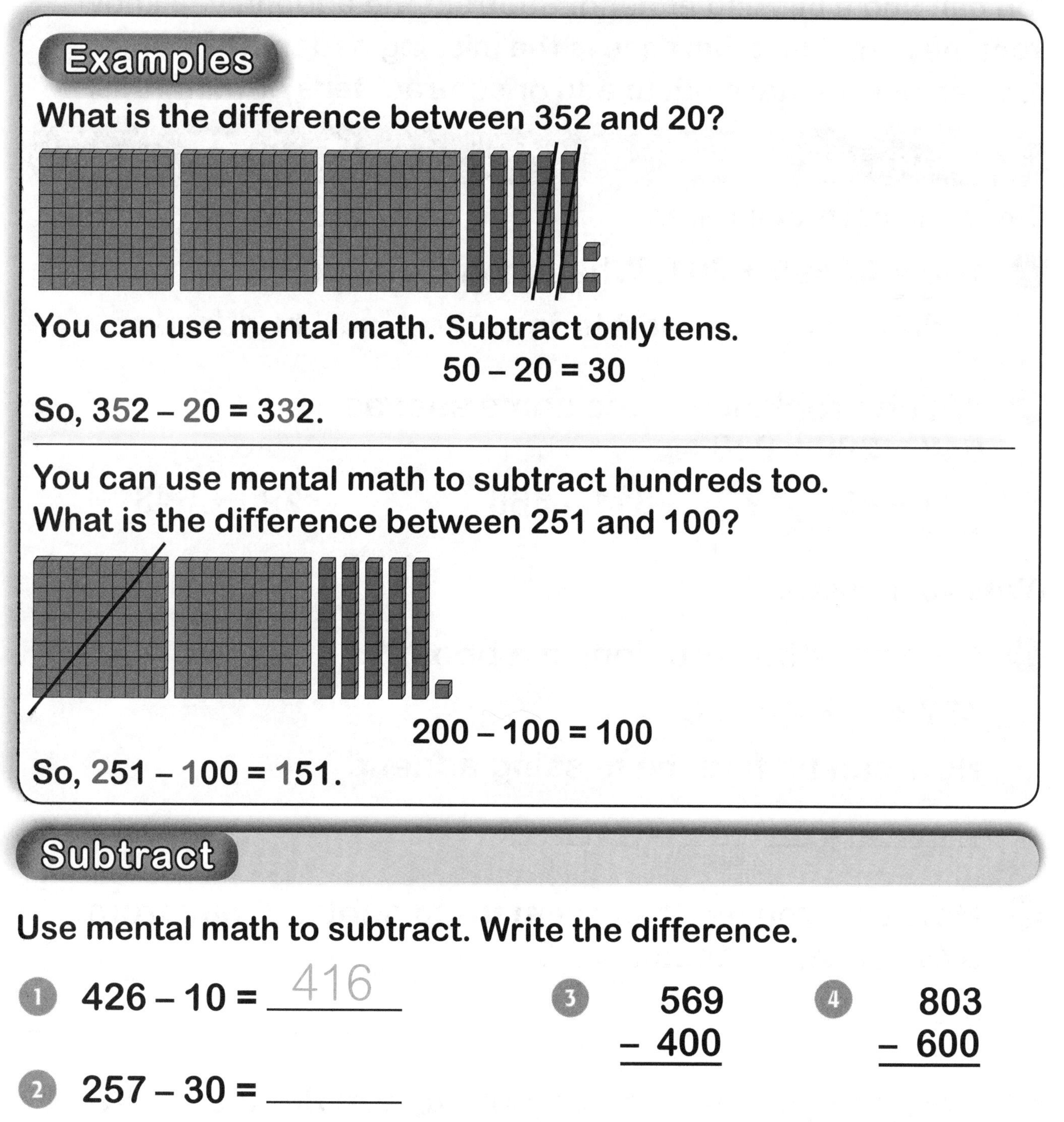

Examples

What is the difference between 352 and 20?

You can use mental math. Subtract only tens.

$50 - 20 = 30$

So, $352 - 20 = 332$.

You can use mental math to subtract hundreds too.
What is the difference between 251 and 100?

$200 - 100 = 100$

So, $251 - 100 = 151$.

Subtract

Use mental math to subtract. Write the difference.

1. $426 - 10 =$ 416
2. $257 - 30 =$ ______
3. $569 - 400$
4. $803 - 600$

Name ______________________

Ways to Add and Subtract

You can add numbers in any order. The sum is the same.
You can group addends in any way. The sum will be the same.
You can find a missing addend. Subtract the addend you know from the sum. The difference is the missing addend.
You can use mental math to add or subtract tens or hundreds.

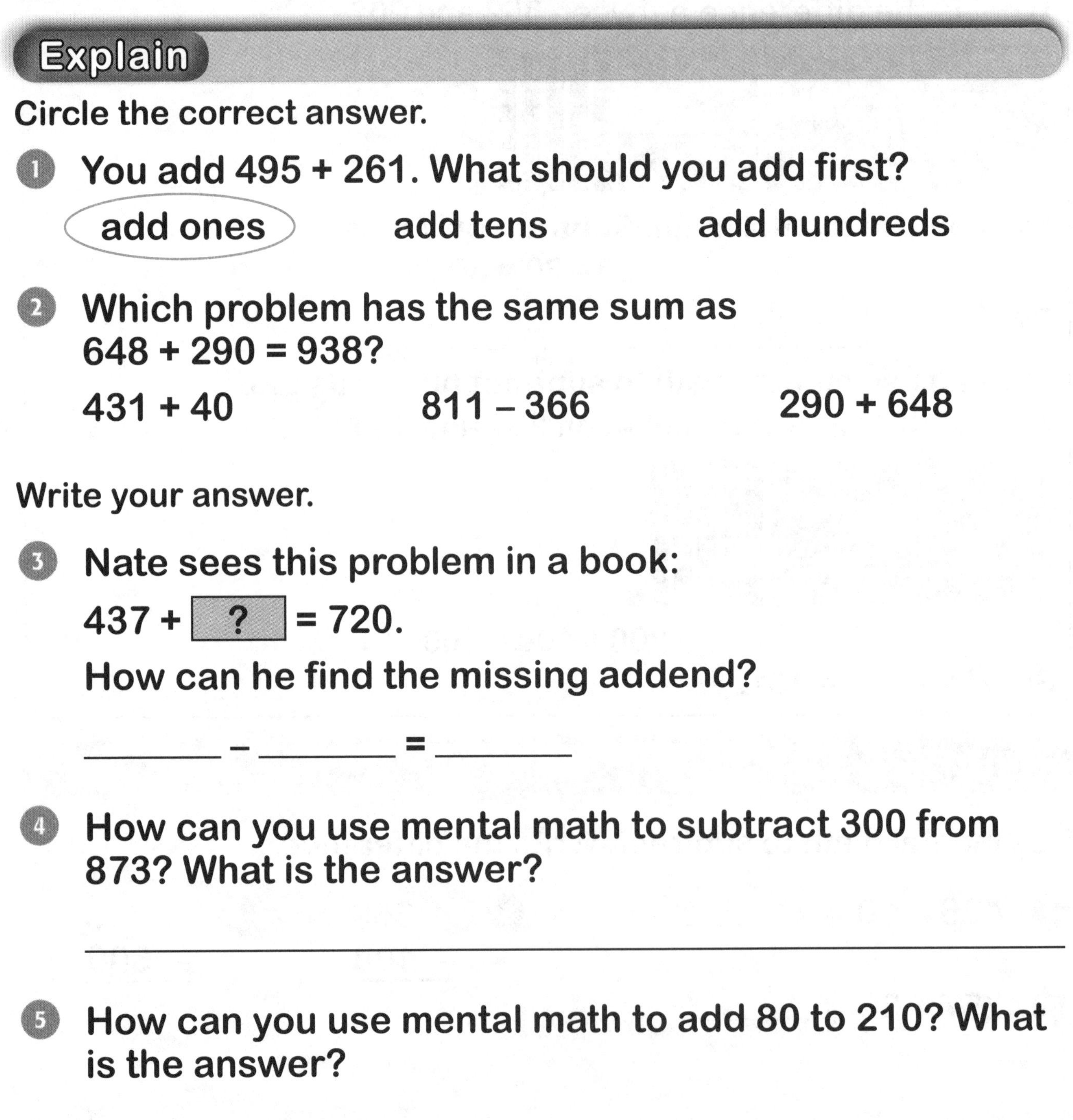

Explain

Circle the correct answer.

1. You add 495 + 261. What should you add first?

   add ones   add tens   add hundreds

2. Which problem has the same sum as 648 + 290 = 938?

   431 + 40   811 – 366   290 + 648

Write your answer.

3. Nate sees this problem in a book:

   437 + [?] = 720.

   How can he find the missing addend?

   ________ – ________ = ________

4. How can you use mental math to subtract 300 from 873? What is the answer?

   ____

5. How can you use mental math to add 80 to 210? What is the answer?

   ____

Name ______________________________

Word Problems Within 1000

Sometimes, you need to find a large sum or a large difference to solve a word problem.

Example

Ty drove 157 miles on one day.

He drove 210 miles the next day.

How many miles did he drive altogether?

$$\begin{array}{r} 157 \text{ addend} \\ +\ 210 \text{ addend} \\ \hline 367 \text{ sum} \end{array}$$

Ty drove 367 miles in all.

Solve

Add or subtract to solve. Write the sum or difference.

1. A store has 450 cans on a shelf.
   It has 322 cans on another shelf.
   How many cans are there in all?

$$\begin{array}{r} 450 \\ +\ 322 \\ \hline 772 \end{array}$$ cans

2. 568 people are at a fair.
   289 go home.
   How many people are still at the fair?

$$\begin{array}{r} 568 \\ -\ 289 \\ \hline \end{array}$$ people

Chapter 5 Test

Name ______________________

Write the sum.

1. 28 + 34 = ______

   34 + 28 = ______

2. 58 + 13 = ______

   13 + 58 = ______

3. 44 + 23 + 11 = ______

   23 + 11 + 44 = ______

4. 228 + 37 + 9 = ______

   37 + 9 + 228 = ______

Use subtraction to find the missing addend. Write the missing addend.

5. 423 + ? = 836

$$\begin{array}{r} 836 \\ -\ 423 \\ \hline \end{array}$$

   423 + ______ = 836

6. 671 + ? = 948

$$\begin{array}{r} 948 \\ -\ 671 \\ \hline \end{array}$$

   671 + ______ = 948

7. 242 + ? = 739

$$\begin{array}{r} 739 \\ -\ 242 \\ \hline \end{array}$$

   242 + ______ = 739

8. 396 + ? = 600

$$\begin{array}{r} 600 \\ -\ 396 \\ \hline \end{array}$$

   396 + ______ = 600

Name ____________________

Add. Write the sum.

9 Add up.

$$\begin{array}{r} 19 \\ 25 \\ 11 \\ +\ 33 \\ \hline \end{array}$$

Add down.

$$\begin{array}{r} 19 \\ 25 \\ 11 \\ +\ 33 \\ \hline \end{array}$$

10 Add up.

$$\begin{array}{r} 27 \\ 15 \\ +\ 28 \\ \hline \end{array}$$

Add down.

$$\begin{array}{r} 28 \\ 27 \\ +\ 15 \\ \hline \end{array}$$

Use mental math to add or subtract. Write the sum or difference.

11

$$\begin{array}{r} 428 \\ +\ 40 \\ \hline \end{array}$$

14

$$\begin{array}{r} 483 \\ -\ 30 \\ \hline \end{array}$$

12

$$\begin{array}{r} 300 \\ +\ 471 \\ \hline \end{array}$$

15

$$\begin{array}{r} 718 \\ -\ 600 \\ \hline \end{array}$$

13

$$\begin{array}{r} 583 \\ +\ 200 \\ \hline \end{array}$$

16

$$\begin{array}{r} 700 \\ -\ 100 \\ \hline \end{array}$$

Solve.

17 A store had 322 shirts.
198 shirts were sold.
How many shirts are left in the store?

$$\begin{array}{r} 322 \\ -\ 198 \\ \hline \end{array}$$

shirts

Name ____________________

Tell and Write Time

The hands on this clock show 10:15.

The hours from midnight to noon are labeled A.M.
If it is 10:15 in the morning, you write 10:15 A.M.

The hours from noon to midnight are labeled P.M.
If it is 10:15 at night, you write 10:15 P.M.

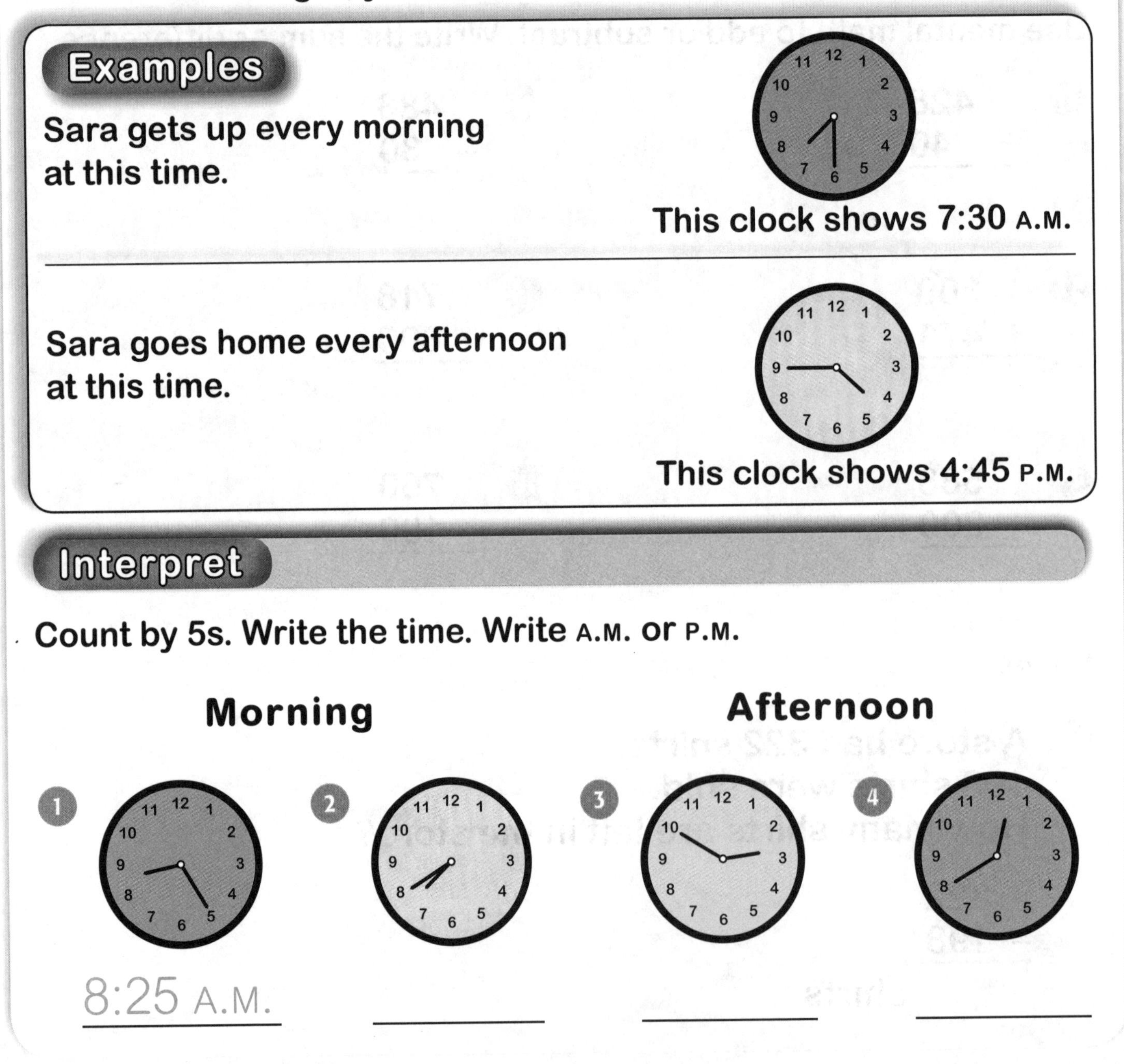

Examples

Sara gets up every morning at this time.

This clock shows 7:30 A.M.

Sara goes home every afternoon at this time.

This clock shows 4:45 P.M.

Interpret

Count by 5s. Write the time. Write A.M. or P.M.

Morning

1. 8:25 A.M.
2. __________

Afternoon

3. __________
4. __________

Name ____________________

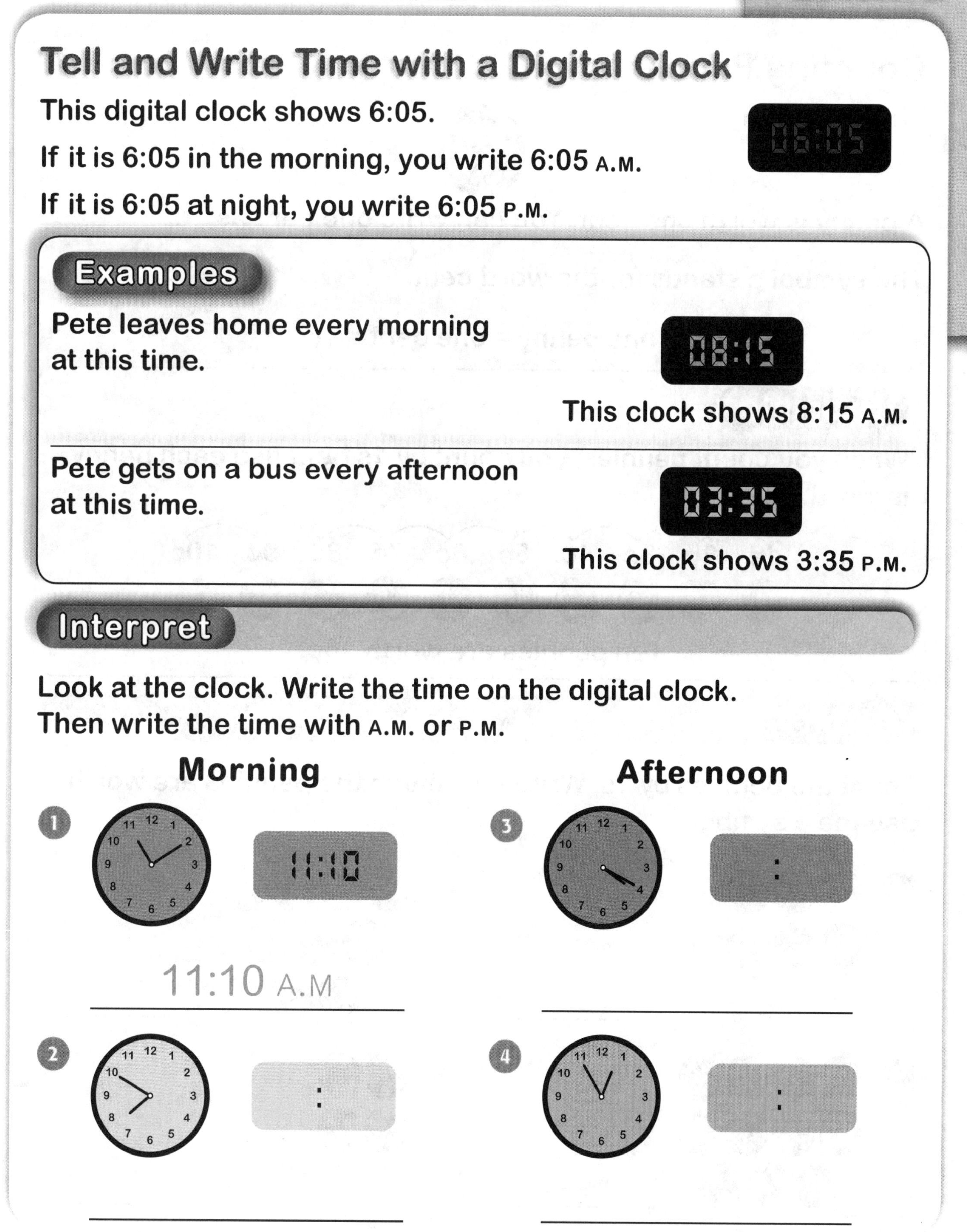

Tell and Write Time with a Digital Clock

This digital clock shows 6:05.

If it is 6:05 in the morning, you write 6:05 A.M.

If it is 6:05 at night, you write 6:05 P.M.

06:05

Examples

Pete leaves home every morning at this time.

08:15

This clock shows 8:15 A.M.

Pete gets on a bus every afternoon at this time.

03:35

This clock shows 3:35 P.M.

Interpret

Look at the clock. Write the time on the digital clock. Then write the time with A.M. or P.M.

Morning

1. 11:10

   11:10 A.M

2. :

   ____

Afternoon

3. :

   ____

4. :

   ____

Name ____________________

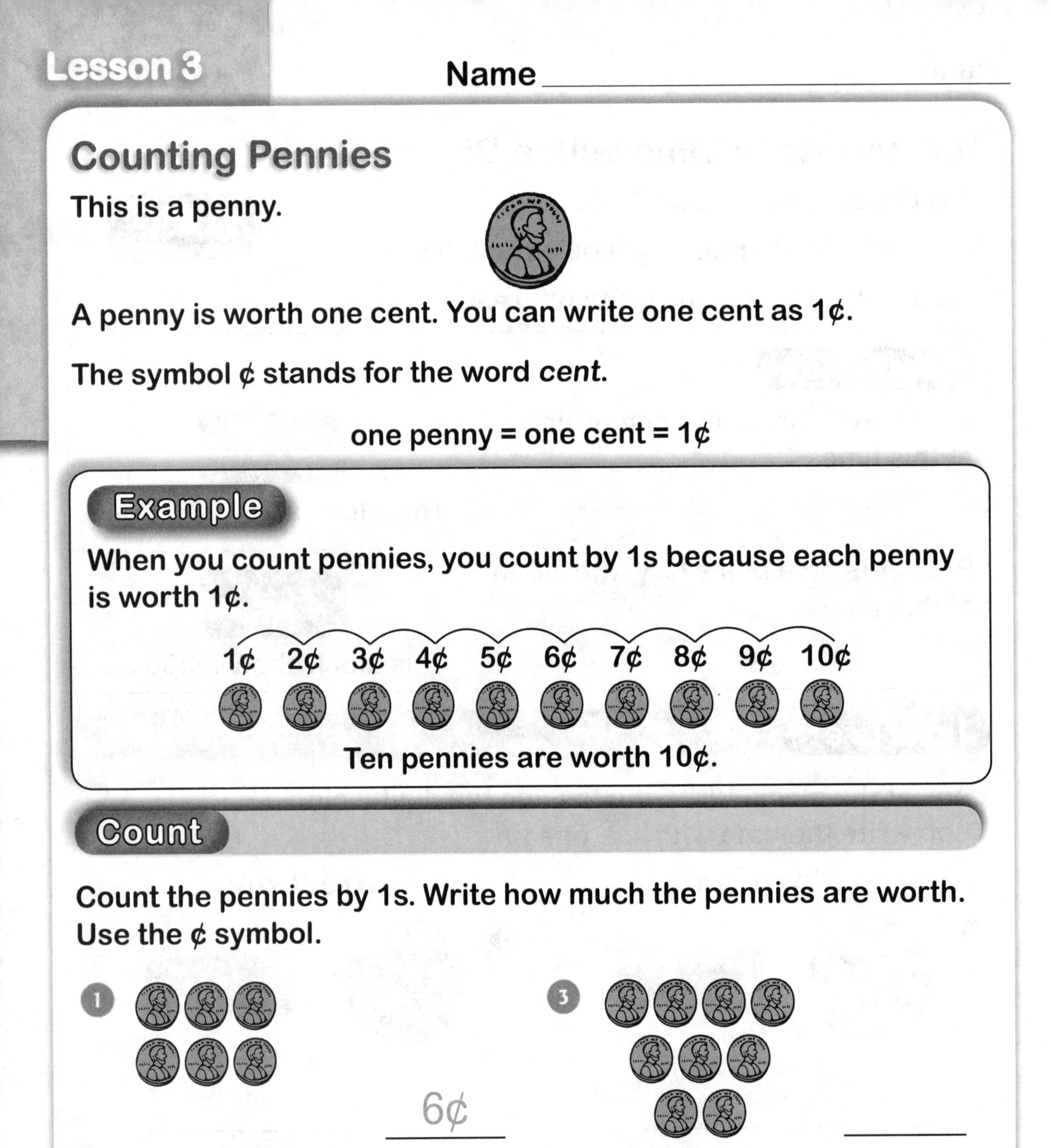

Counting Pennies

This is a penny.

A penny is worth one cent. You can write one cent as 1¢.

The symbol ¢ stands for the word *cent*.

one penny = one cent = 1¢

Example

When you count pennies, you count by 1s because each penny is worth 1¢.

1¢ 2¢ 3¢ 4¢ 5¢ 6¢ 7¢ 8¢ 9¢ 10¢

Ten pennies are worth 10¢.

Count

Count the pennies by 1s. Write how much the pennies are worth. Use the ¢ symbol.

1. 6¢

2. ________

3. ________

4. ________

Name ______________________________

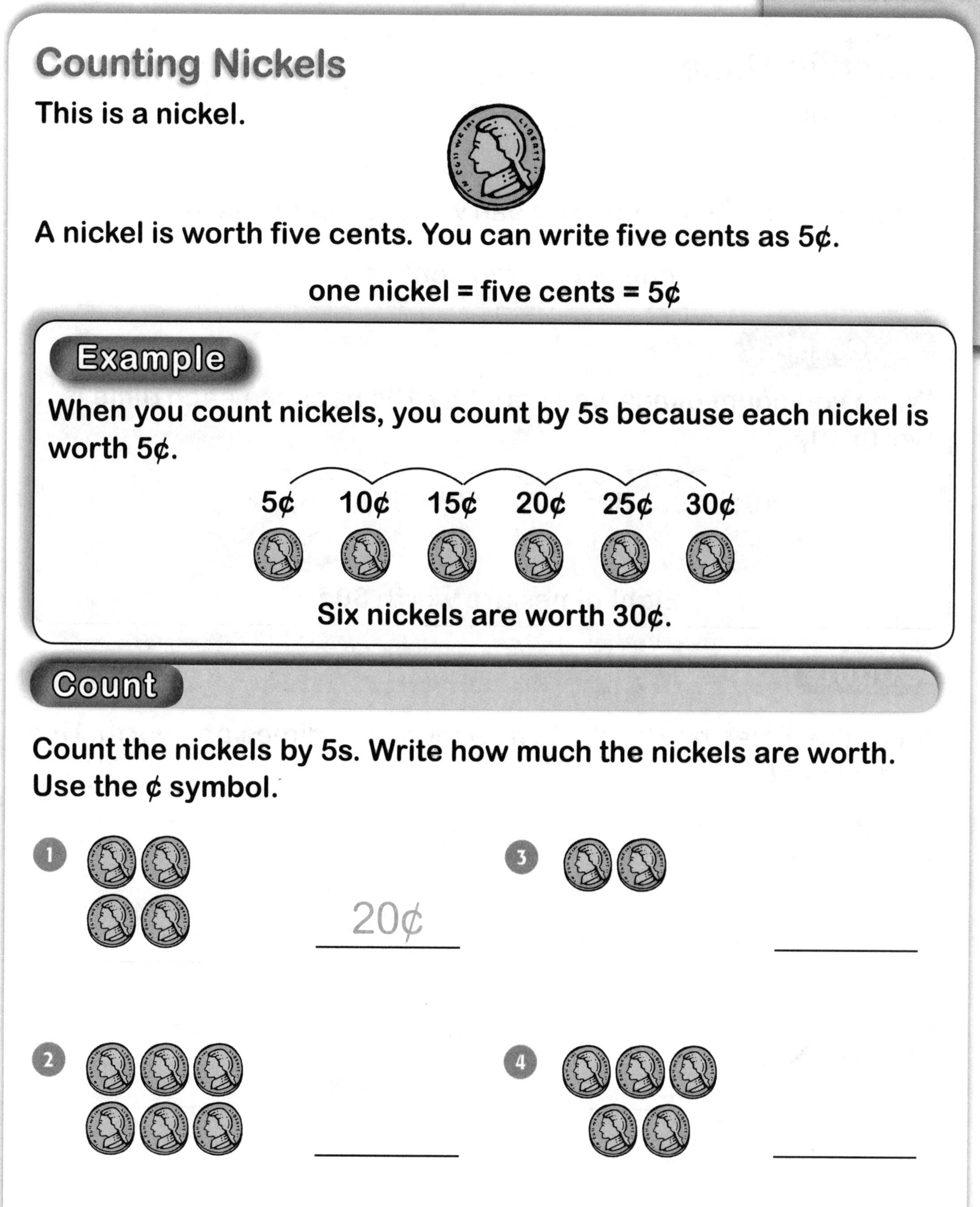

Counting Nickels

This is a nickel.

A nickel is worth five cents. You can write five cents as 5¢.

one nickel = five cents = 5¢

Example

When you count nickels, you count by 5s because each nickel is worth 5¢.

5¢ 10¢ 15¢ 20¢ 25¢ 30¢

Six nickels are worth 30¢.

Count

Count the nickels by 5s. Write how much the nickels are worth. Use the ¢ symbol.

1. 20¢

2. ________

3. ________

4. ________

Name ____________________

Counting Dimes

This is a dime.

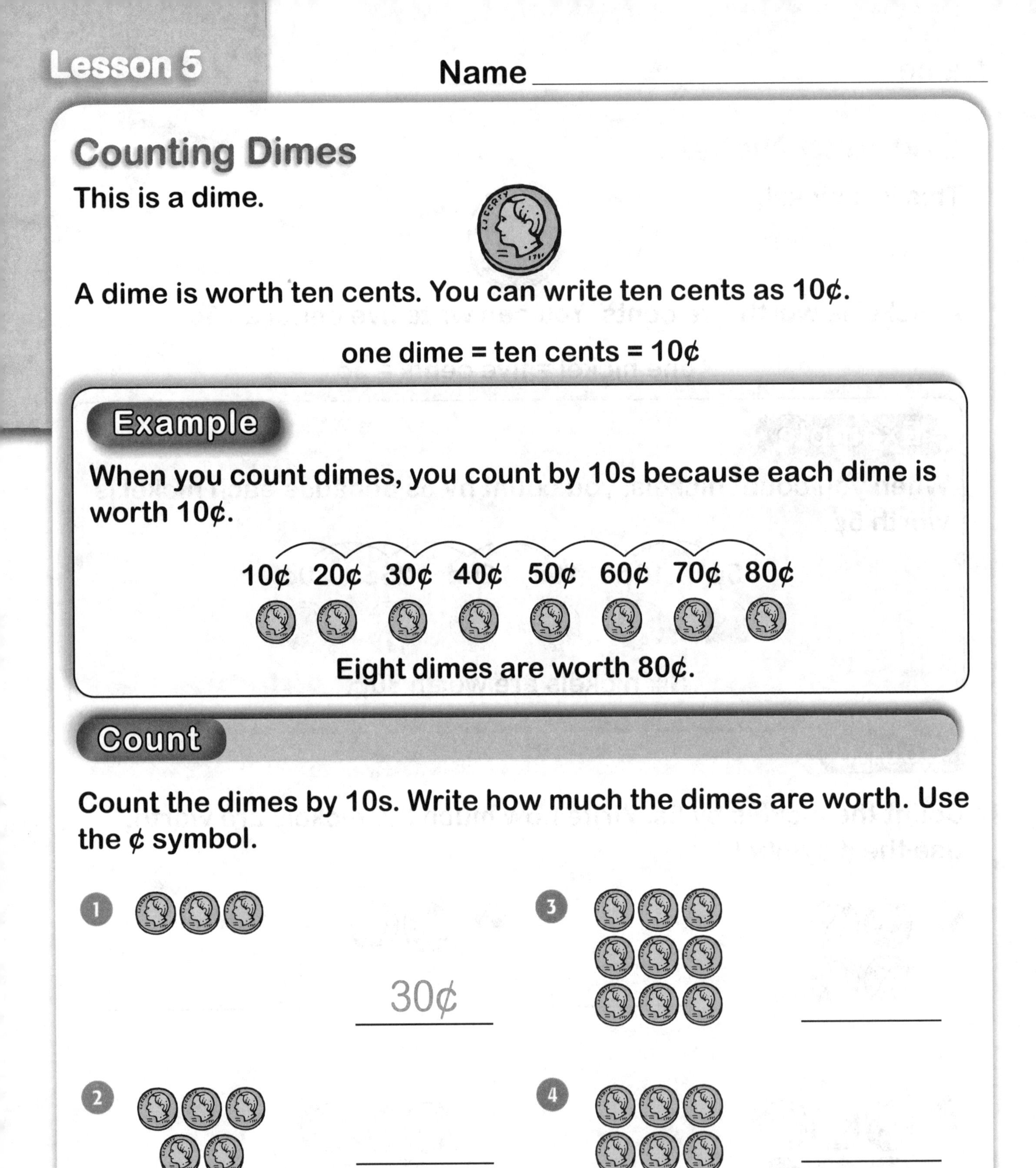

A dime is worth ten cents. You can write ten cents as 10¢.

one dime = ten cents = 10¢

Example

When you count dimes, you count by 10s because each dime is worth 10¢.

10¢ 20¢ 30¢ 40¢ 50¢ 60¢ 70¢ 80¢

Eight dimes are worth 80¢.

Count

Count the dimes by 10s. Write how much the dimes are worth. Use the ¢ symbol.

1. 30¢

2. ________

3. ________

4. ________

Name ______________________________

Counting Pennies, Nickels, and Dimes

A penny is worth 1¢. A nickel is worth 5¢. A dime is worth 10¢.

Examples

Count pennies by 1s.	1¢	2¢	3¢	4¢	5¢
Count nickels by 5s.	5¢	10¢	15¢	20¢	25¢
Count dimes by 10s.	10¢	20¢	30¢	40¢	50¢

Count

Count the dimes, nickels, and pennies by 10s, 5s, and 1s. Write the total. Use the ¢ symbol.

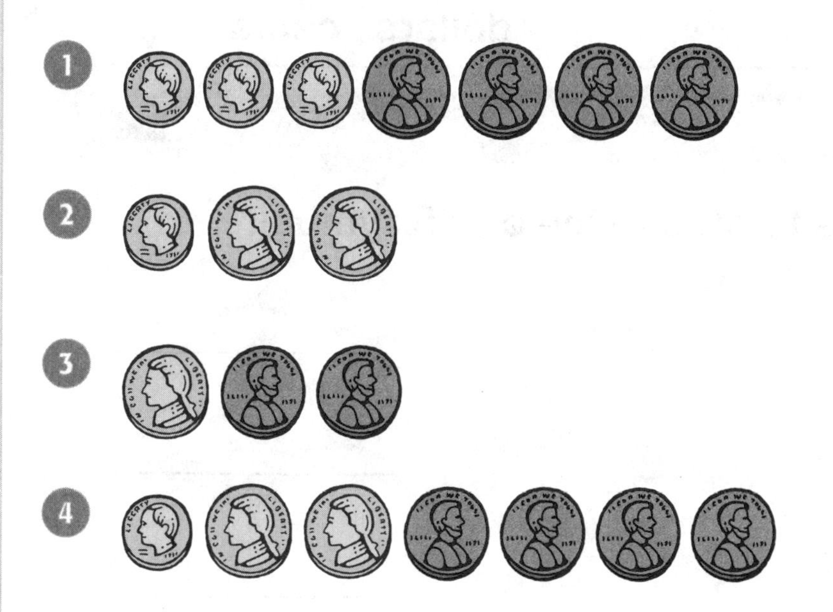

1. 34¢
2. ______
3. ______
4. ______

Name ____________________

Counting Quarters and Dollars

This is a quarter.

A quarter is worth twenty-five cents. You can write twenty-five cents as 25¢.

This is a one-dollar bill.

A dollar is worth one hundred cents. You can write 100¢ as $1.00. The $ symbol stands for the word *dollar*.

Examples

one quarter = twenty-five cents = 25¢

one dollar = one hundred cents = 100¢ = $1.00

one dollar and one quarter = $1.00 and 25¢ = $1.25

↑ dollars ↑ cents

Count

Count the money. Write the total. Use the ¢ or $ symbol.

1. 75¢
2. ____________
3. ____________
4. ____________

Name ______________________

More Practice Counting Money

A dollar is worth 100¢ or $1.00.

A quarter is worth 25¢.

A dime is worth 10¢.

A nickel is worth 5¢.

A penny is worth 1¢.

Example

If you have one dollar bill, one quarter, one dime, one nickel, and one penny in your pocket, how much money do you have?

one dollar	=	$1.00
one quarter	=	25¢
one dime	=	10¢
one nickel	=	5¢
one penny	=	1¢
Total:		$1.41

Count

Count the money. Write the total. Use the ¢ or $ symbol.

1. $1.32

2. ______

3. ______

Lesson 9

Name ______________________

Word Problems with Money

You can use what you know about money to solve word problems.

Example

One apple costs 25¢. Ana has two dimes and five pennies. Can Ana buy an apple?

Think:

10¢ 20¢ 21¢ 22¢ 23¢ 24¢ 25¢

An apple costs 25¢. Ana has 25¢. Yes, Ana can buy an apple.

Solve

Write how much money the person has.
Write how much money the person needs. Use the ¢ or $ symbol.
Then write yes or no to answer the question.

1. One plum costs 14¢. John has a quarter. He wants to buy two plums.

   Has 25¢

   Needs 28¢

   Can John buy two plums? no

2. Eva has two quarters, five nickels, and four pennies. One bunch of grapes costs 79¢.

   Has ______

   Needs ______

   Can Eva buy a bunch of grapes? ______

Name ______________________

Write the time. Write A.M. or P.M.

Morning

1 

______

Afternoon

2 

______

Look at the clock. Write the time on the digital clock. Then write the time with A.M. or P.M.

Morning

3  

______

Afternoon

4  

______

Count the pennies, nickels, or dimes. Write how much they are worth. Use the ¢ symbol.

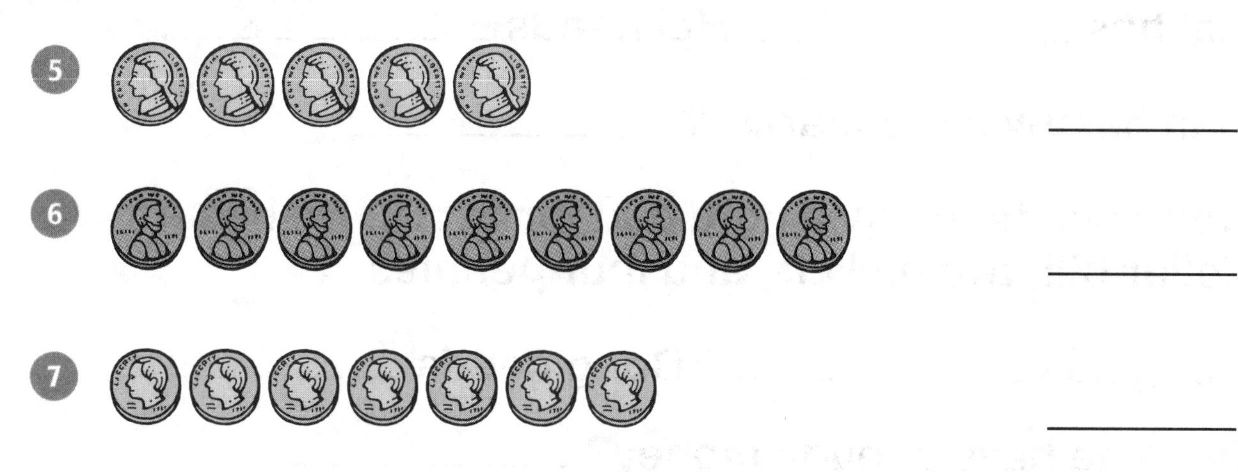

5 ______

6 ______

7 ______

Name ____________________

Count the dimes, nickels, and pennies by 10s, 5s, and 1s. Write the total. Use the ¢ symbol.

8 ____________

9 ____________

Count the money. Write the total. Use the ¢ or $ symbol.

10 ____________

11 ____________

12 ____________

Write how much money the person has. Write how much money the person needs. Use the ¢ or $ symbol. Then answer the question. Write yes or no.

13 One red marker costs 55¢. Hal has one quarter, two dimes, and two nickels.

Hal has ____________. Hal needs ____________.

Can he buy a red marker? ____________

14 Doug wants to buy a note pad for $1.19. He has a dollar bill, two nickels, and four pennies.

Doug has ____________. Doug needs ____________.

Does he have enough money? ____________

Name ______________________________

Lesson 1

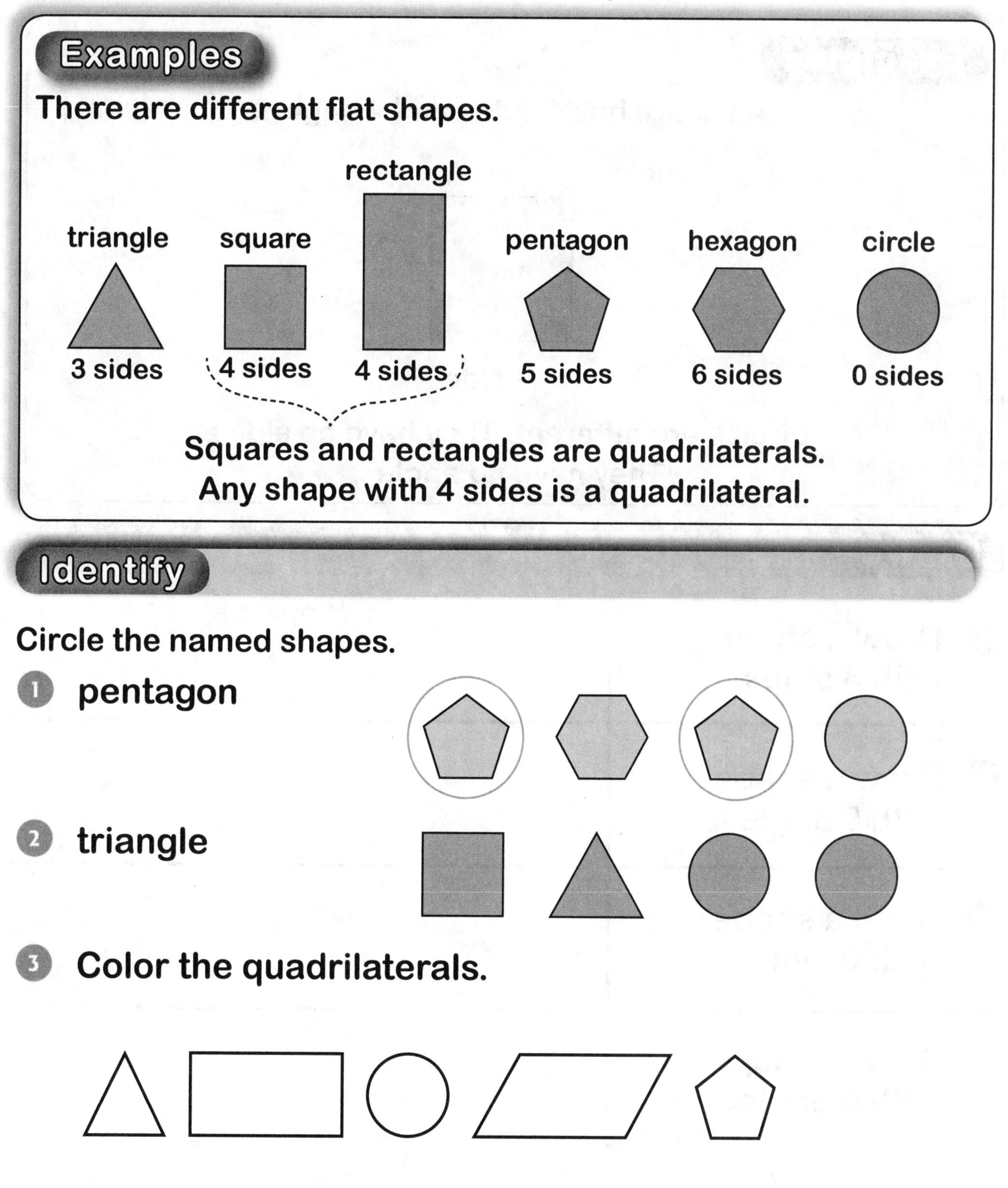

Identifying Shapes

Some shapes are flat shapes. Each shape has a name.

Examples

There are different flat shapes.

Squares and rectangles are quadrilaterals.
Any shape with 4 sides is a quadrilateral.

Identify

Circle the named shapes.

1. pentagon
2. triangle
3. Color the quadrilaterals.

Lesson 2

Name ____________________

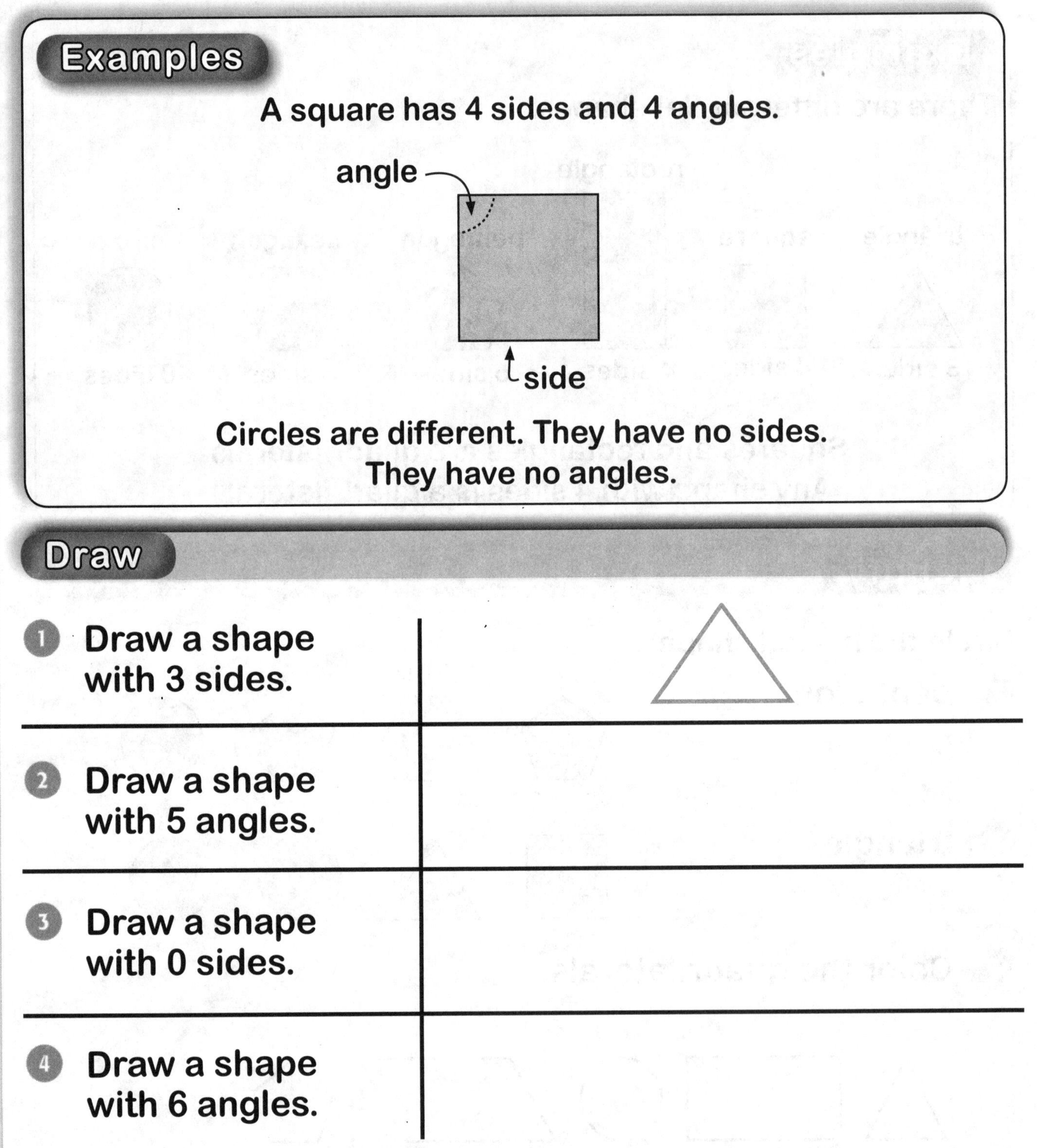

Drawing Shapes

Most shapes have sides and angles.

Examples

A square has 4 sides and 4 angles.

Circles are different. They have no sides.
They have no angles.

Draw

1. **Draw a shape with 3 sides.**	
2. **Draw a shape with 5 angles.**	
3. **Draw a shape with 0 sides.**	
4. **Draw a shape with 6 angles.**	

Name ______________________

Drawing Solid Shapes

Shapes that are not flat are solid shapes. Each shape has a name.

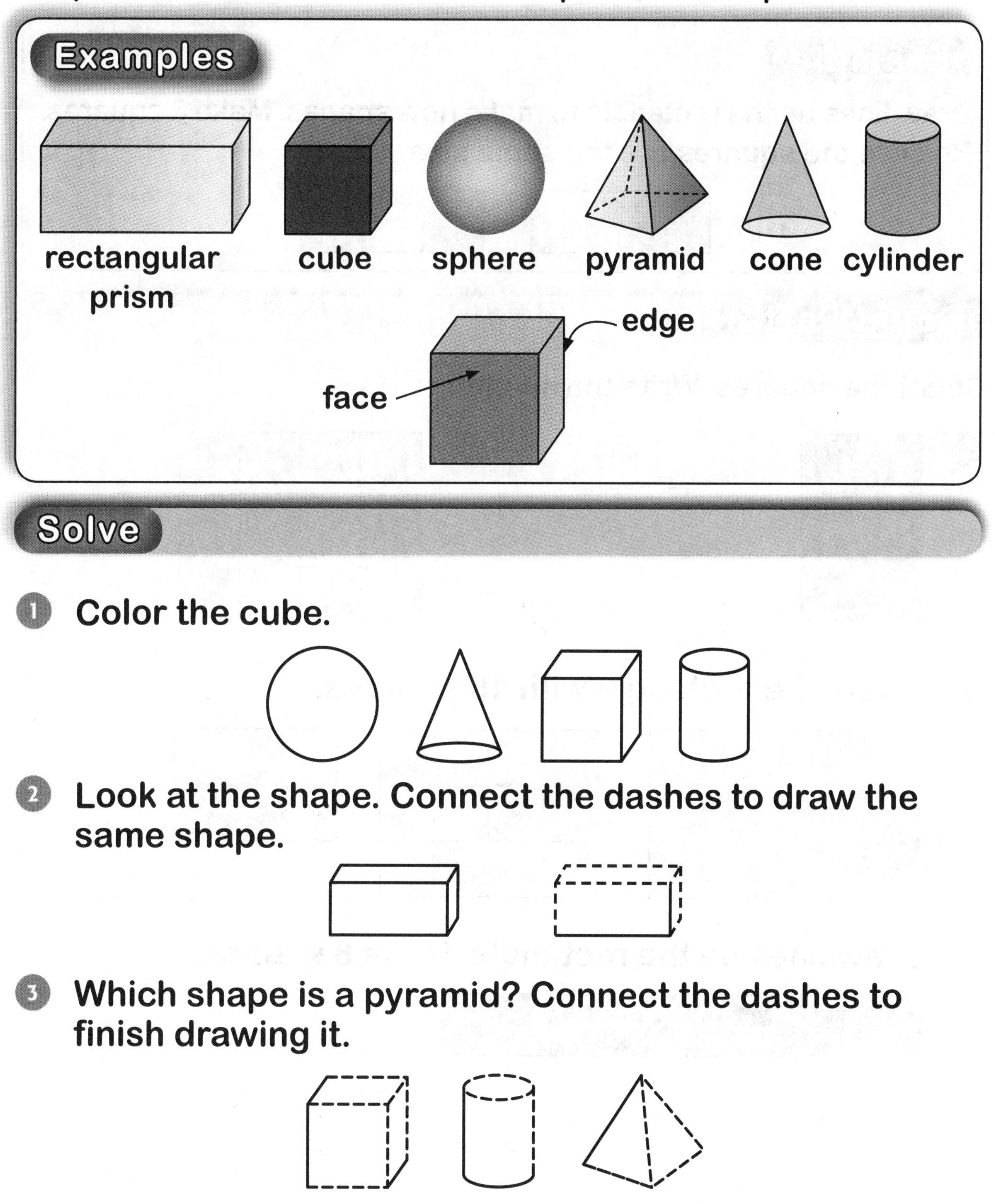

1. Color the cube.

2. Look at the shape. Connect the dashes to draw the same shape.

3. Which shape is a pyramid? Connect the dashes to finish drawing it.

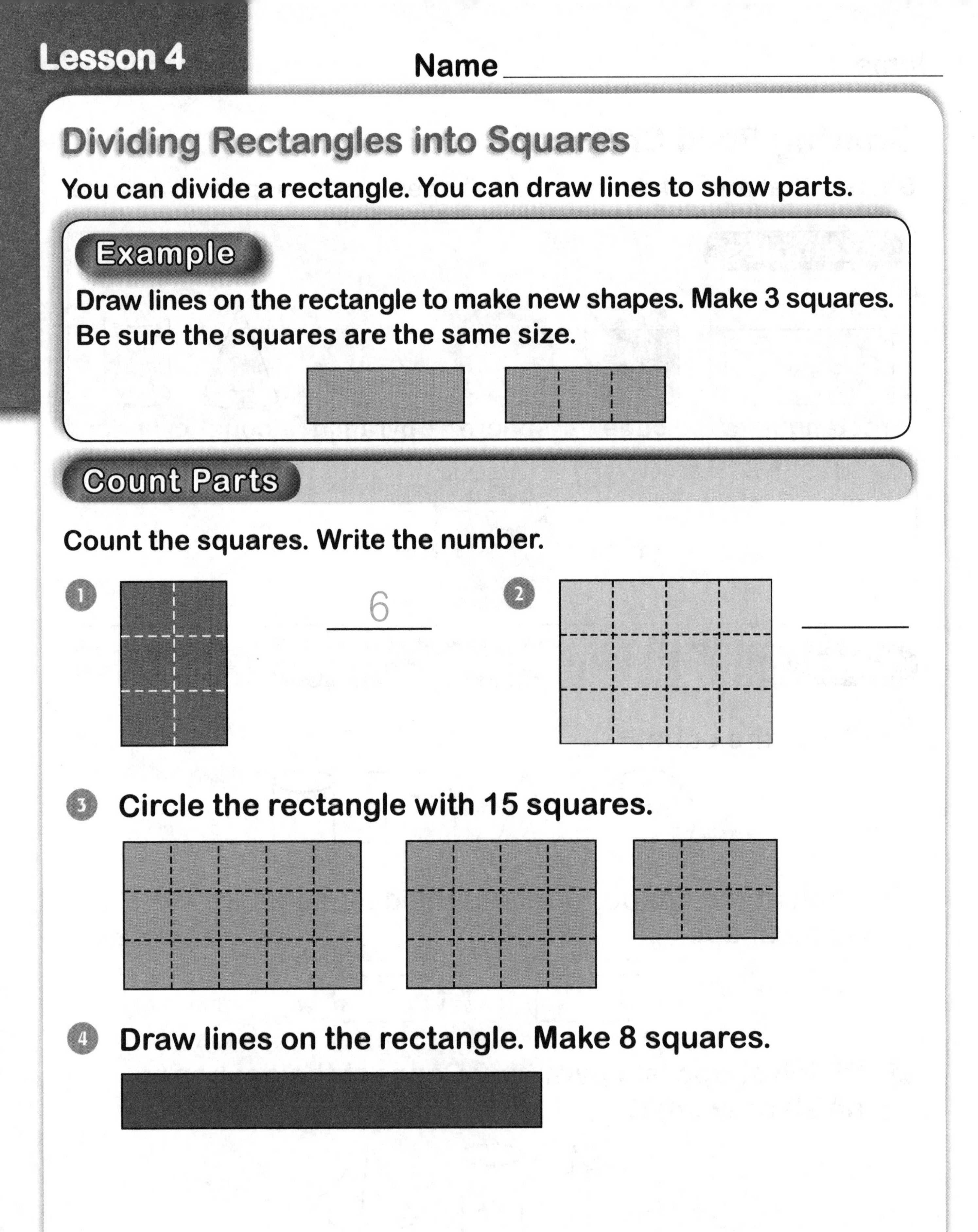

Lesson 4

Name ______________________

Dividing Rectangles into Squares

You can divide a rectangle. You can draw lines to show parts.

Example

Draw lines on the rectangle to make new shapes. Make 3 squares. Be sure the squares are the same size.

Count Parts

Count the squares. Write the number.

1. 6

2. ______

3. Circle the rectangle with 15 squares.

4. Draw lines on the rectangle. Make 8 squares.

Name ____________________

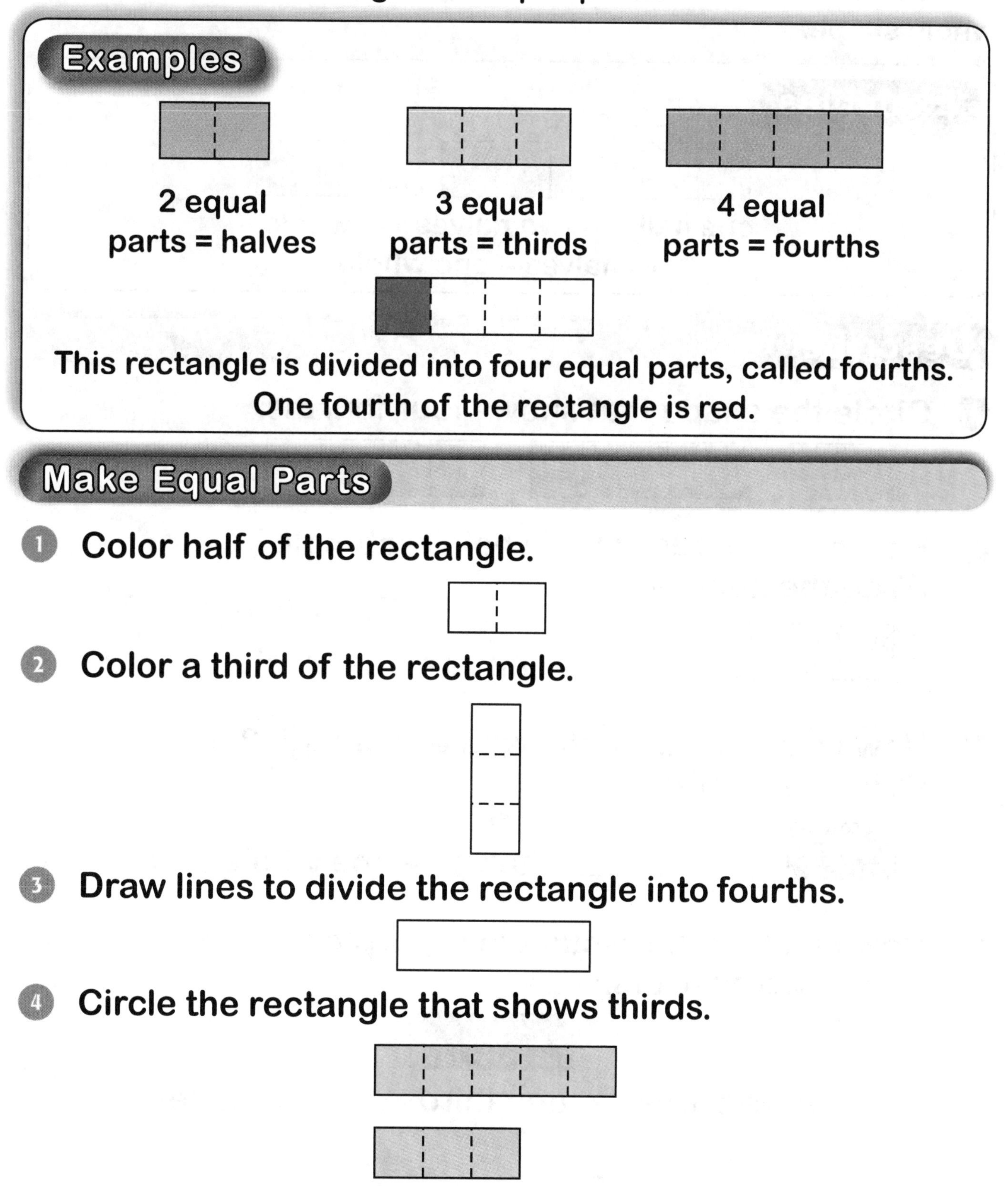

Dividing Rectangles into Equal Parts

You can divide a rectangle into equal parts.

Examples

2 equal parts = halves

3 equal parts = thirds

4 equal parts = fourths

This rectangle is divided into four equal parts, called fourths. One fourth of the rectangle is red.

Make Equal Parts

1. Color half of the rectangle.
2. Color a third of the rectangle.
3. Draw lines to divide the rectangle into fourths.
4. Circle the rectangle that shows thirds.

Name ____________________

Describing Whole Rectangles

You can divide a shape into parts. Together, the parts make a whole shape.

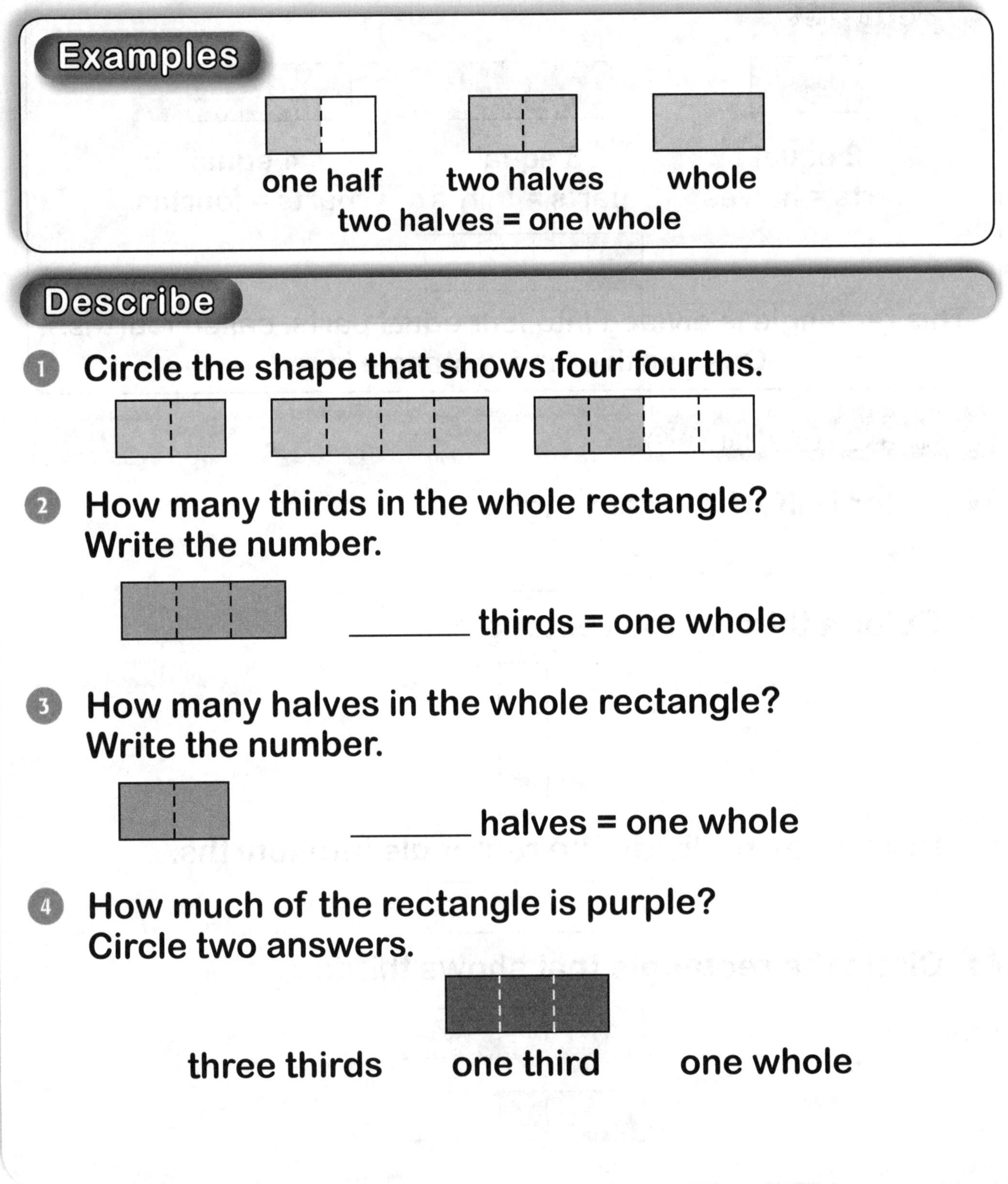

Examples

one half   two halves   whole

two halves = one whole

Describe

1. Circle the shape that shows four fourths.

2. How many thirds in the whole rectangle? Write the number.

   ______ thirds = one whole

3. How many halves in the whole rectangle? Write the number.

   ______ halves = one whole

4. How much of the rectangle is purple? Circle two answers.

   three thirds   one third   one whole

Name ______________________________

Identifying Equal Parts of Circles

You can divide circles into halves, thirds, and fourths.

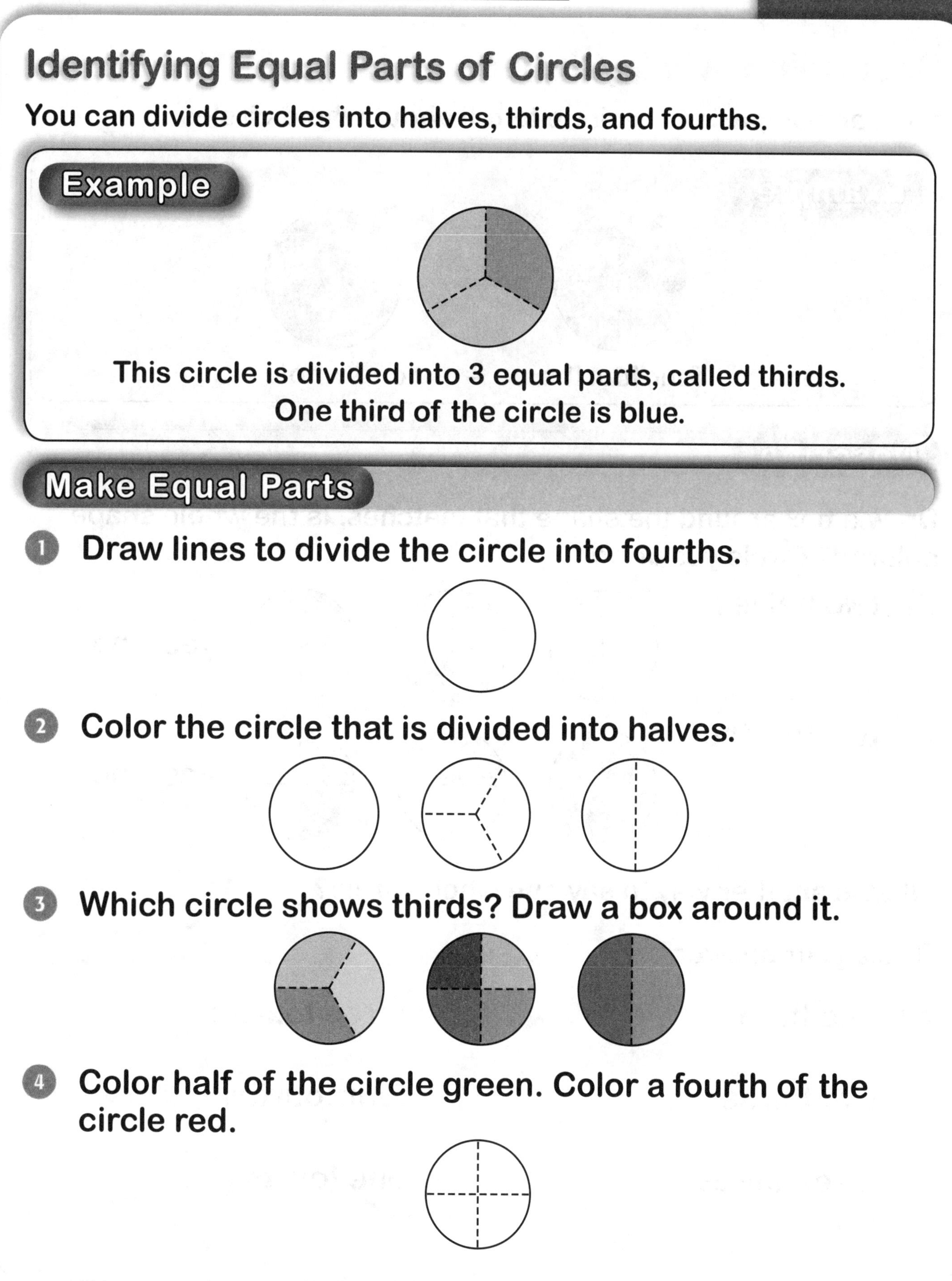

Example

This circle is divided into 3 equal parts, called thirds. One third of the circle is blue.

Make Equal Parts

1. Draw lines to divide the circle into fourths.

2. Color the circle that is divided into halves.

3. Which circle shows thirds? Draw a box around it.

4. Color half of the circle green. Color a fourth of the circle red.

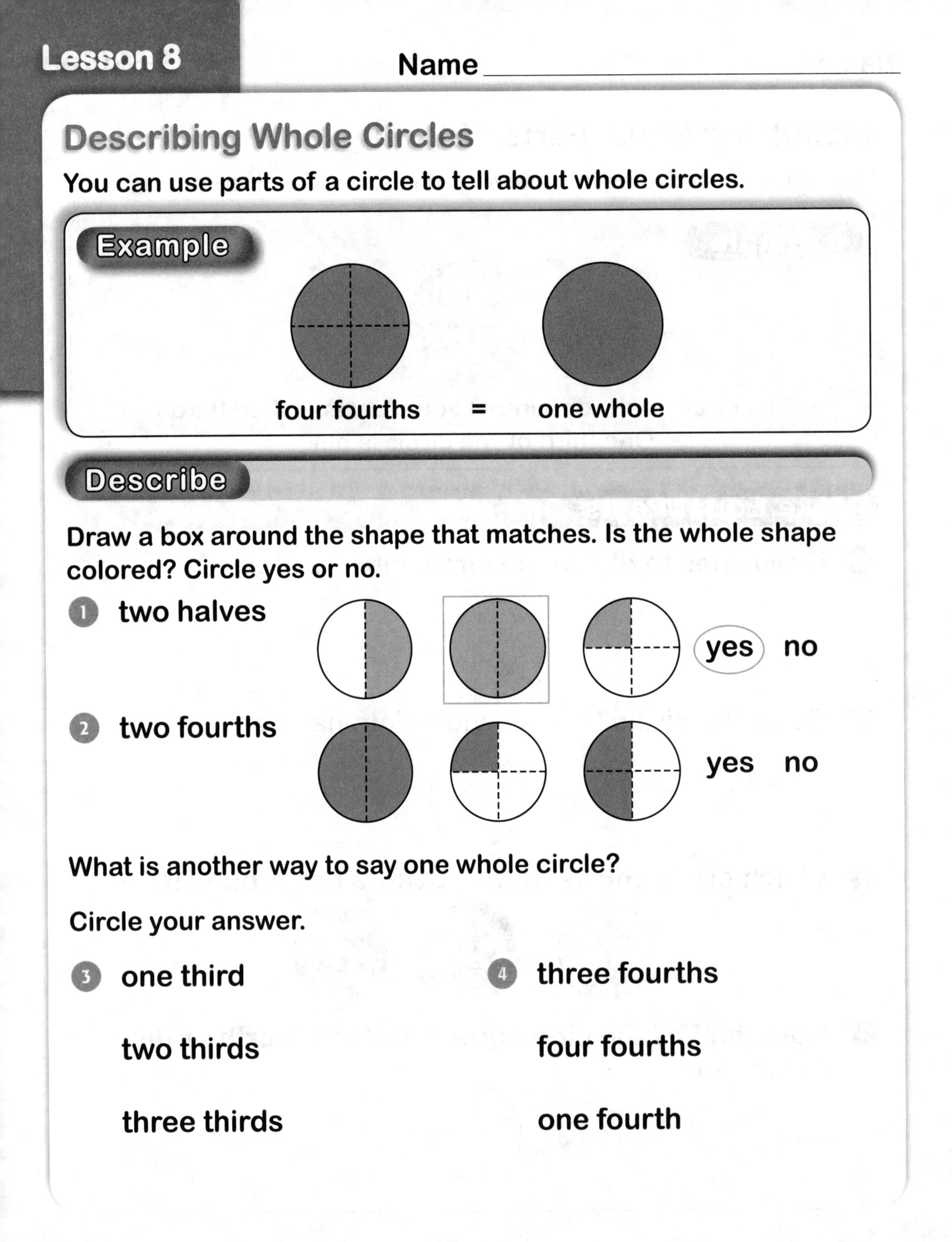

Lesson 8

Name ______________________

Describing Whole Circles

You can use parts of a circle to tell about whole circles.

Example

four fourths = one whole

Describe

Draw a box around the shape that matches. Is the whole shape colored? Circle yes or no.

1. two halves — yes no

2. two fourths — yes no

What is another way to say one whole circle?

Circle your answer.

3. one third
   two thirds
   three thirds

4. three fourths
   four fourths
   one fourth

Name ______________________

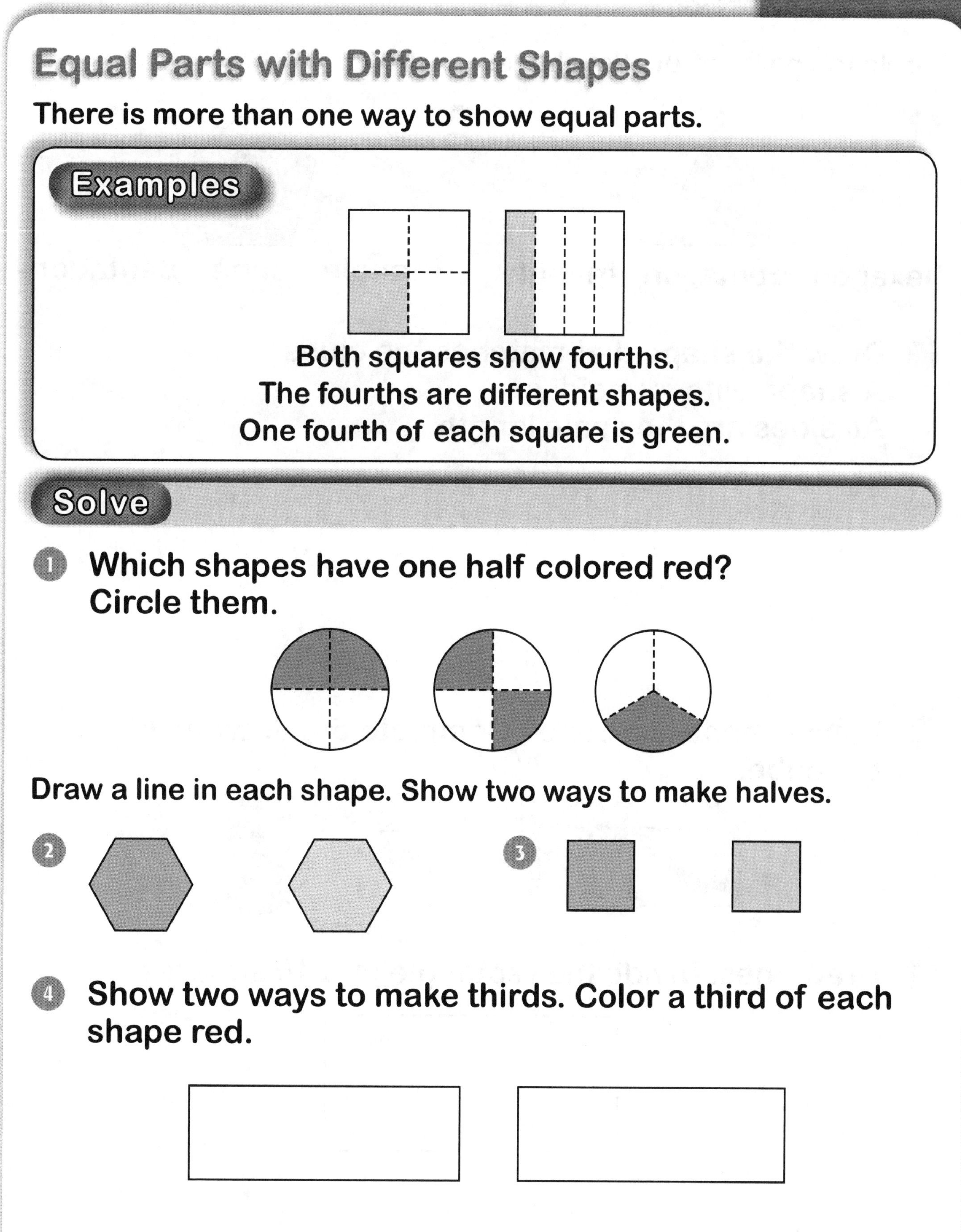

Equal Parts with Different Shapes

There is more than one way to show equal parts.

Examples

Both squares show fourths.
The fourths are different shapes.
One fourth of each square is green.

Solve

1. Which shapes have one half colored red? Circle them.

Draw a line in each shape. Show two ways to make halves.

2.

3.

4. Show two ways to make thirds. Color a third of each shape red.

Chapter 7 Test

Name ____________________________

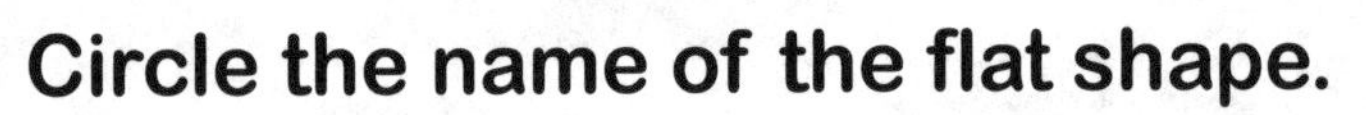

Circle the name of the flat shape.

1 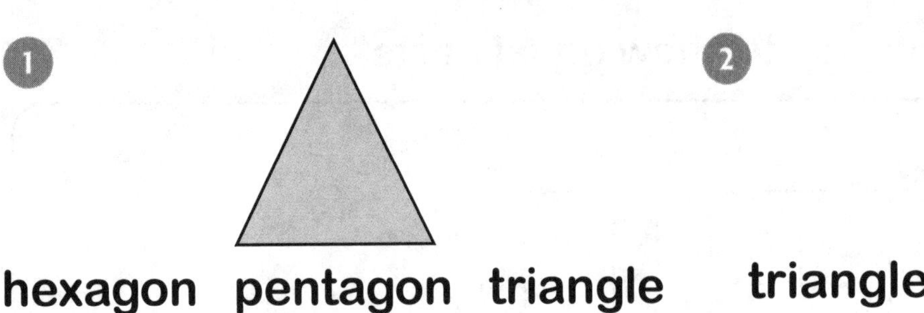

hexagon pentagon triangle

2 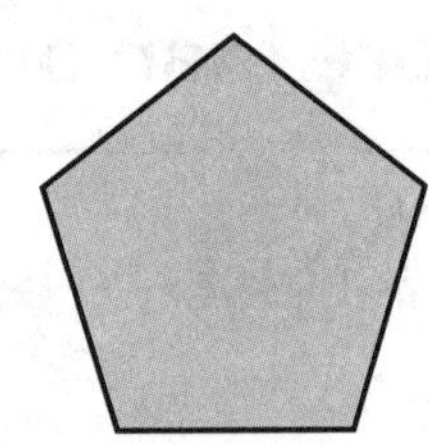

triangle cube pentagon

3 Draw the shape that matches the clues.
A shape with four sides.
All sides are the same length.

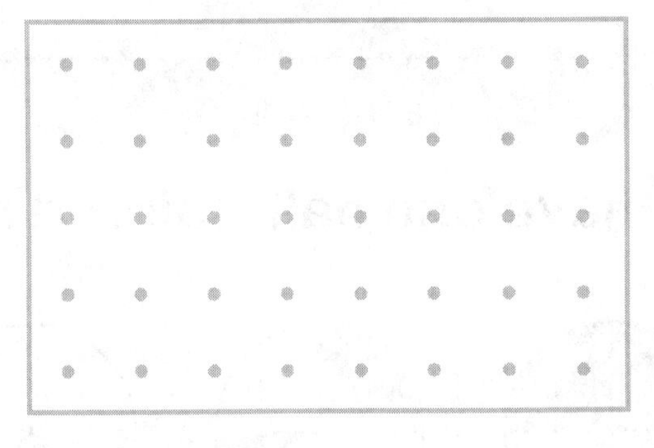

4 Which shape is a cube? Complete the drawing of the cube.

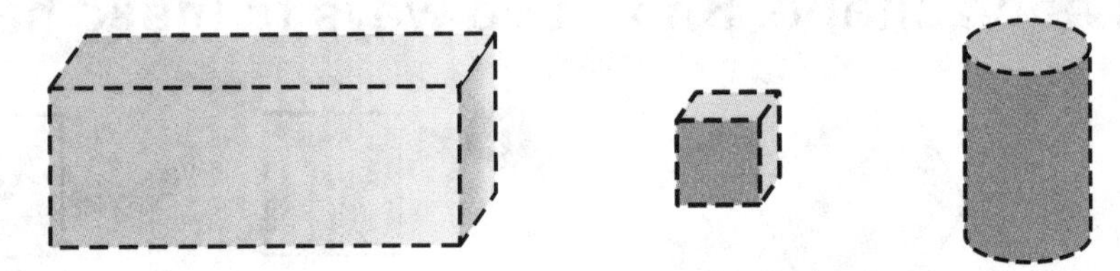

5 Draw lines. Divide the rectangle into 10 squares.

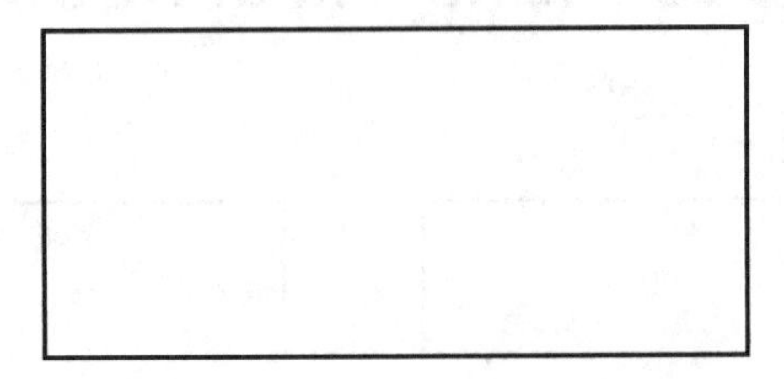

Name ______________________

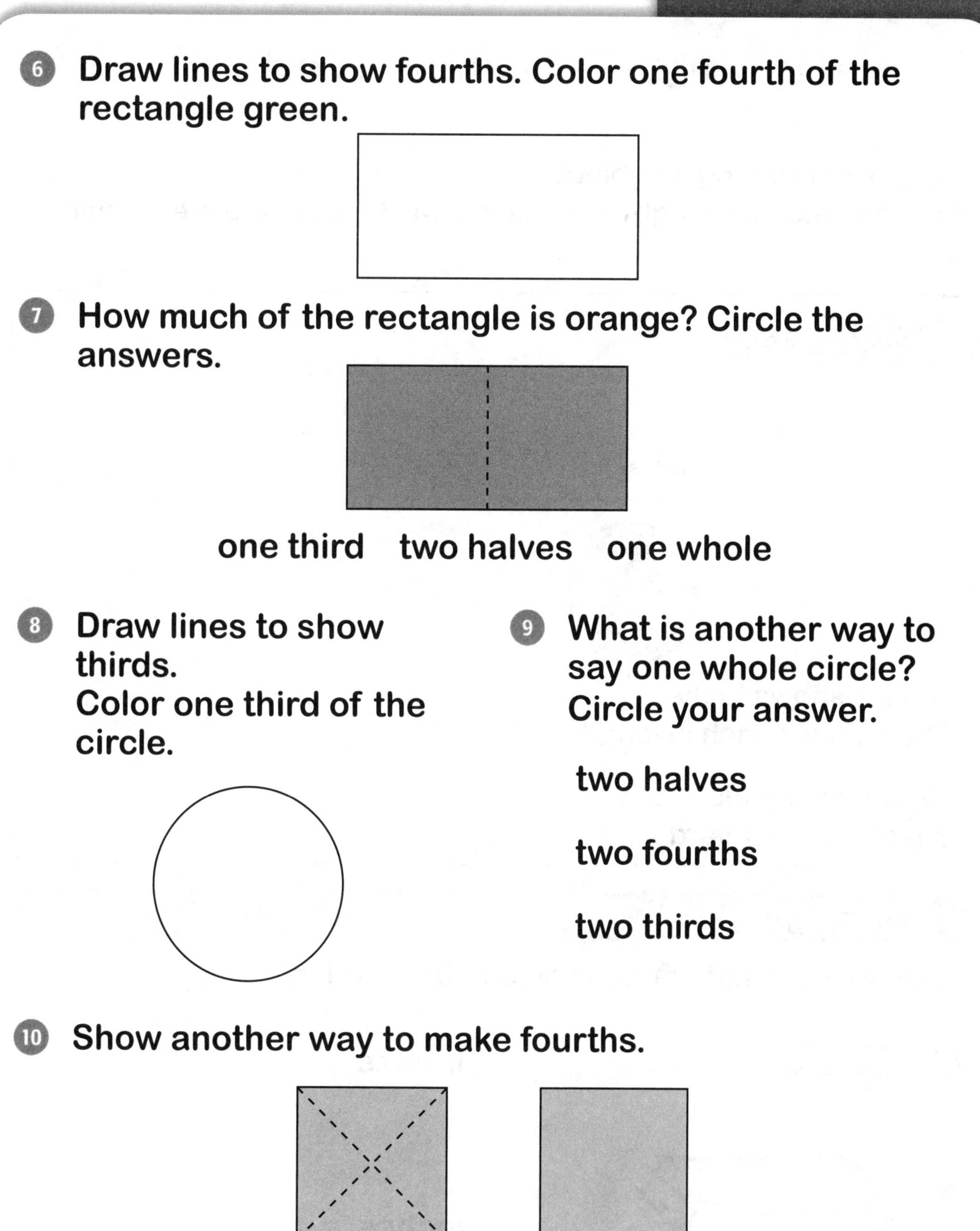

6 Draw lines to show fourths. Color one fourth of the rectangle green.

7 How much of the rectangle is orange? Circle the answers.

one third   two halves   one whole

8 Draw lines to show thirds.
Color one third of the circle.

9 What is another way to say one whole circle? Circle your answer.

two halves

two fourths

two thirds

10 Show another way to make fourths.

Name ____________________

Measuring Length Using Inches, Feet, and Yards

Length is how long an object is.
You can measure length with rulers, yardsticks, and measuring tapes.

Examples

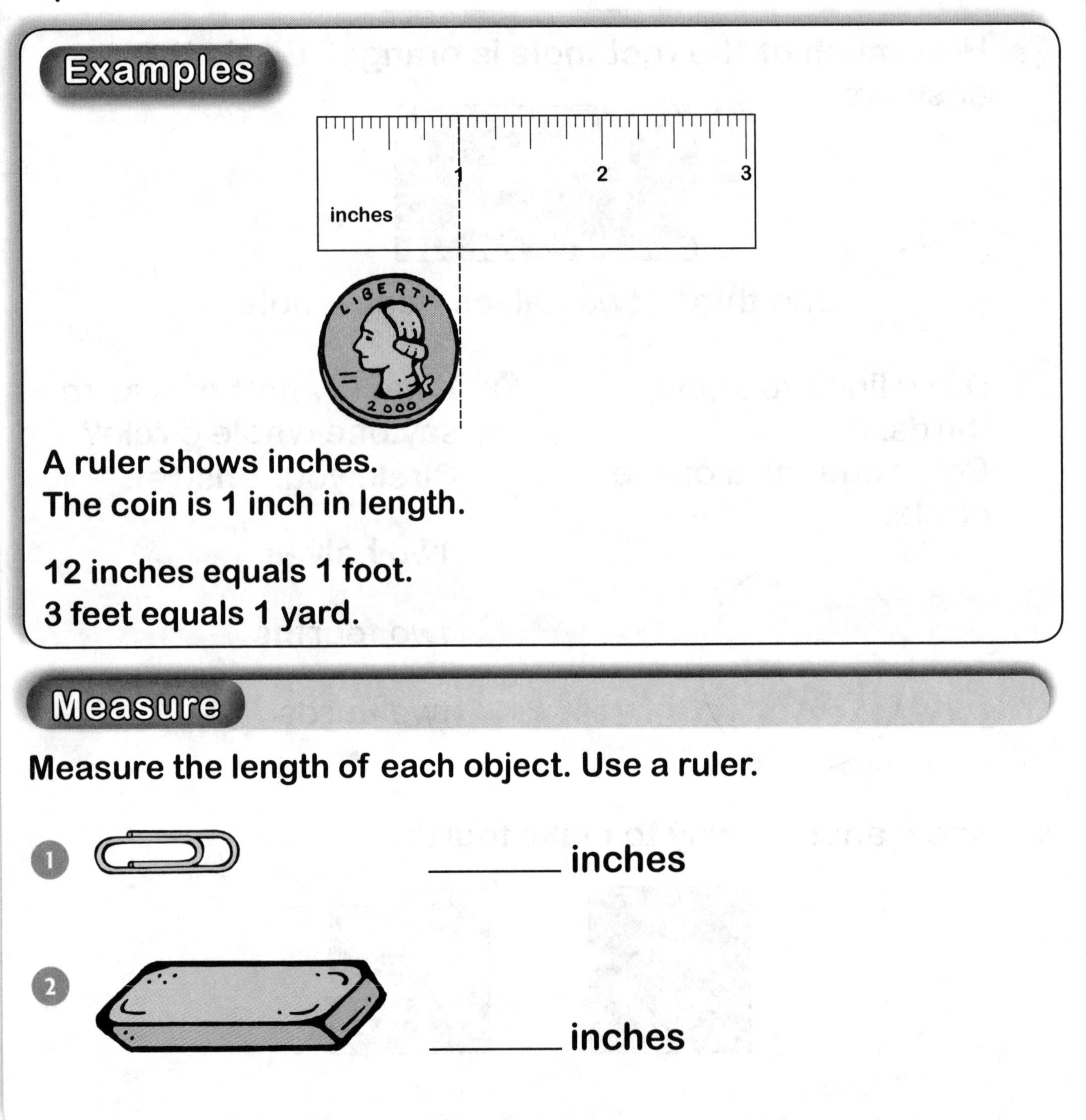

A ruler shows inches.
The coin is 1 inch in length.

12 inches equals 1 foot.
3 feet equals 1 yard.

Measure

Measure the length of each object. Use a ruler.

1. ________ inches

2. ________ inches

Name ______________________

Measuring Length Using Centimeters and Meters

You can also measure length using centimeters and meters.

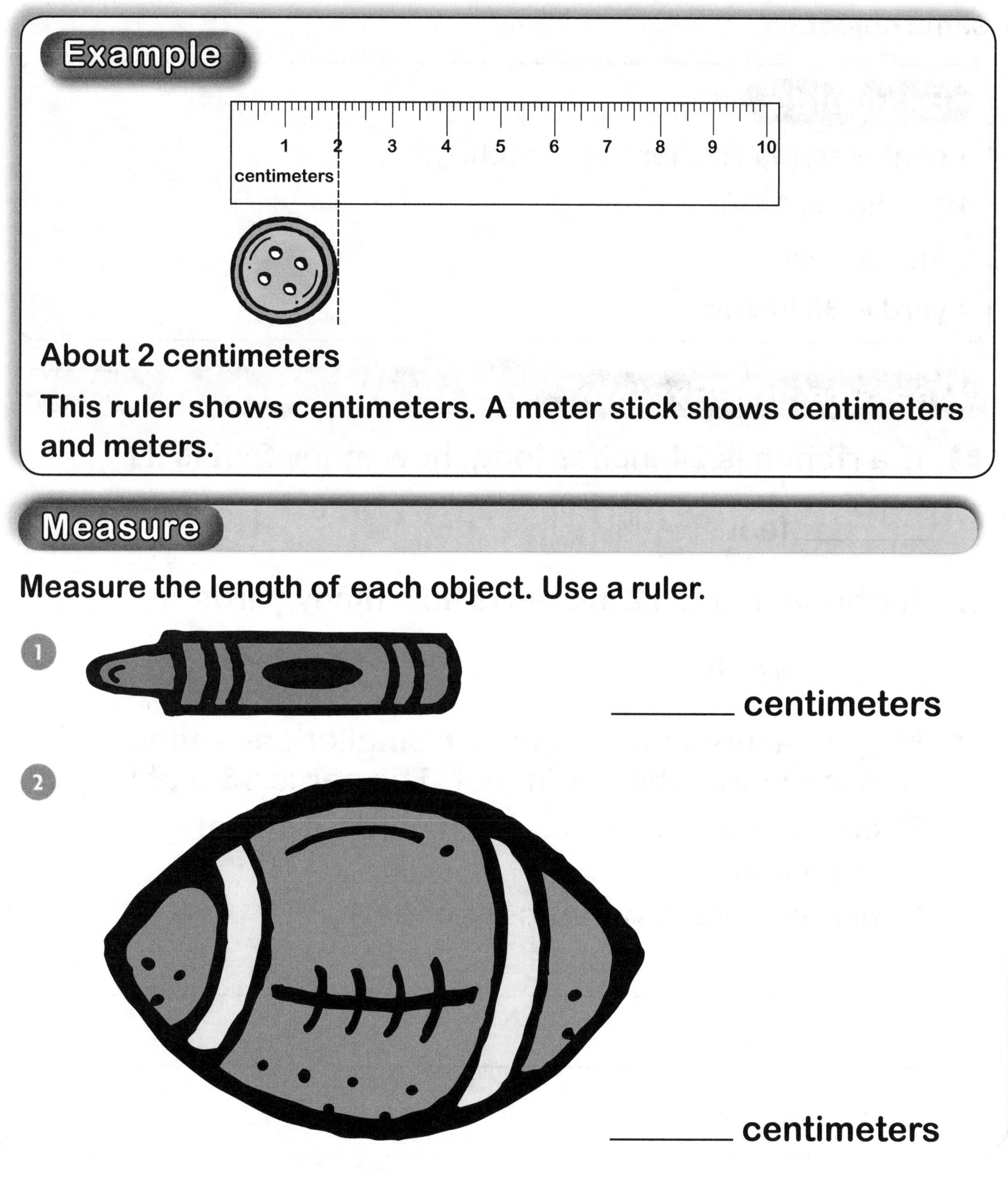

Example

About 2 centimeters

This ruler shows centimeters. A meter stick shows centimeters and meters.

Measure

Measure the length of each object. Use a ruler.

1. ______ centimeters

2. ______ centimeters

Lesson 3

Name ____________________

Measuring the Same Length Using Inches, Feet, and Yards

You can use different units to measure the length of the same object.

Examples

1 centimeter is shorter than 1 inch.

12 inches is 1 foot.

3 feet is 1 yard

1 yard is 36 inches.

Measure and Explain

1. If a ribbon is 24 inches long, how many feet is it?

   __2__ feet

2. A ribbon that is 72 inches is how many yards?

   ______ yards

3. Meg measures the length of a smaller car 2 times. She measures the car in feet. The car is 15 feet long.
   Then she measures the car in yards. The car is 5 yards long.
   Why are there less yards than feet?

   ____

   ____

Name ______________________

Measuring the Same Length Using Centimeters and Meters

Example

You can measure length in centimeters and meters.
One meter is equal to 100 centimeters.

Measure and Explain

1. If an object is 4 meters in length, how many centimeters is it? 400 centimeters

2. How many meters long is an object that is 900 centimeters long? ________ meters

3. If an object is 8 meters in length, how many centimeters is it? ________ centimeters

4. How many meters long is an object that is 200 centimeters long? ________ meters

5. A volleyball net is 11 meters long.
The net is 1,100 centimeters long.
Why are there more centimeters than meters?

____

____

Name ______________________

Estimating Length Using Inches and Feet

You can estimate length. An estimate is a careful guess.

Example

Dean has a large paperclip. He wants to know how long it is.

He guesses the length of the paperclip. He estimates its length as best as he can.
He thinks it is about 2 inches long.

Then Dean measures it with a ruler to check his guess.

The paperclip is 1 inch long.

Estimate and Measure

Estimate the length of the object. Then measure the length. Ask an adult to help you measure.

1.

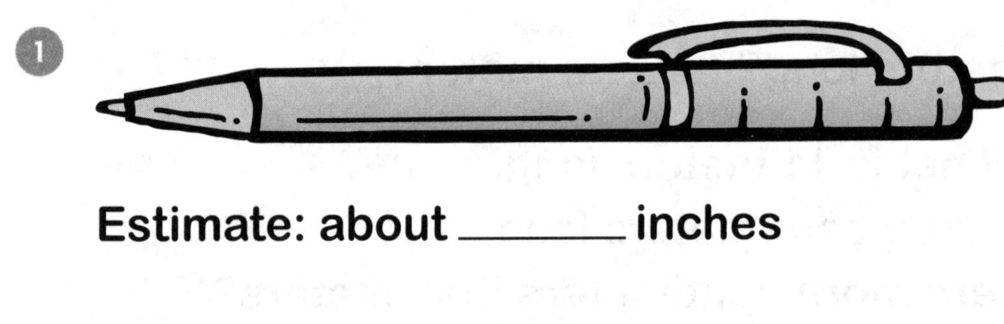

Estimate: about ______ inches

Measure: ______ inches

Name ______________________________

Estimating Length Using Centimeters and Meters

You can estimate length using centimeters and meters.

Example

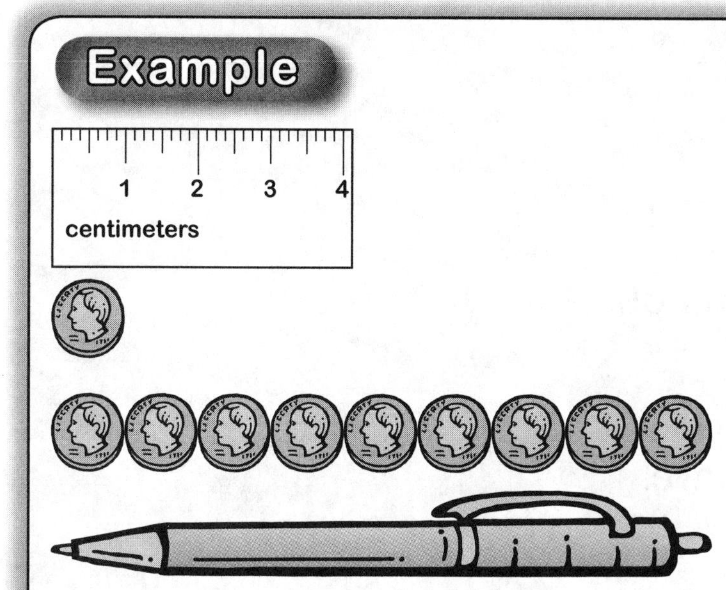

The pen is about 9 coins long.

Since 1 coin is about 1 centimeter, the pen is about 9 centimeters long

Estimate and Measure

Estimate the length of the object. Then measure the length. Ask an adult to help you measure.

1. Estimate: about ________ centimeters

   Measure: ________ centimeters

Lesson 7

Name ______________________

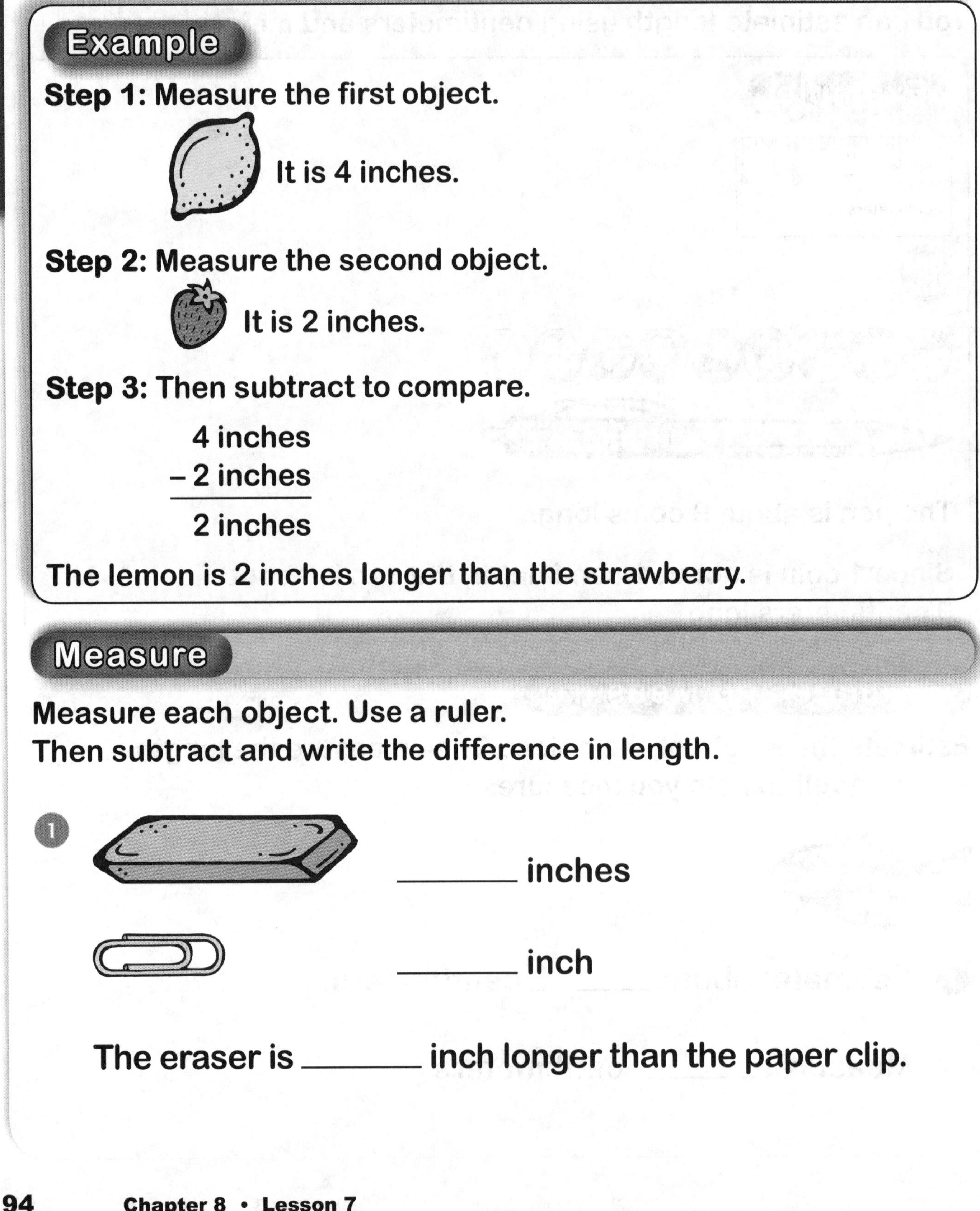

Measuring to Compare

You can measure objects to compare their lengths.

Example

Step 1: Measure the first object.

It is 4 inches.

Step 2: Measure the second object.

It is 2 inches.

Step 3: Then subtract to compare.

$$\begin{array}{r} 4 \text{ inches} \\ -\ 2 \text{ inches} \\ \hline 2 \text{ inches} \end{array}$$

The lemon is 2 inches longer than the strawberry.

Measure

Measure each object. Use a ruler.
Then subtract and write the difference in length.

1. ______ inches

______ inch

The eraser is ______ inch longer than the paper clip.

Name ______________________

Adding and Subtracting Lengths

You can add and subtract lengths to solve problems.

Example

Mr. Gray has two pieces of wood.

They measure 5 feet long when placed together.

One piece is 2 feet long.

How long is the other piece?

2 feet + [?] = 5 feet

5 – 2 = 3

The other piece is 3 feet long.

Solve

Write the equations to find the answer.

1. Greg has 16 inches of string. He uses 9 inches to wrap a present. How many inches of string are left?

   16 – 9 = 7 inches

2. Allison walks 20 feet. Then she walks 12 more feet. How many feet did Allison walk in all?

   ____

Name ______________________________

Using a Number Line to Show Lengths

Example

You can use a number line to show the lengths of objects.

The numbers on the number line can stand for a unit of measure.

0 1 2 3 4 5 6 7 8 9 10 11 12 13 14 15 16 17 18 19 20

A sheet of paper is 11 inches long.
You can put a dot above the 11 on the number line.
This shows you measured one object, and it is 11 inches long.

Measure and Record

Use a ruler. Measure these objects in inches.
Put dots on the number line to show their lengths.

0 1 2 3 4 5 6 7 8 9 10

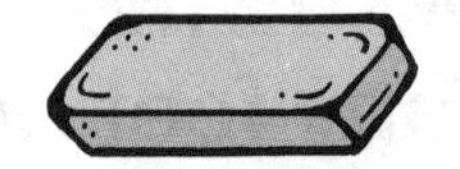

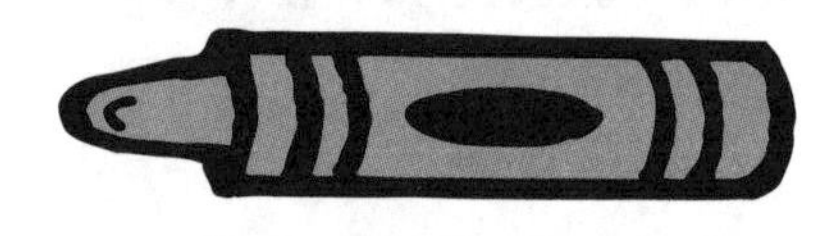

1. How many objects were 1 inch long? ________

2. How many objects were 2 inches long? ________

Name ____________________

Adding and Subtracting Using a Number Line

You can use number lines to help you add or subtract.
When you add, you count on.
When you subtract, you count back.

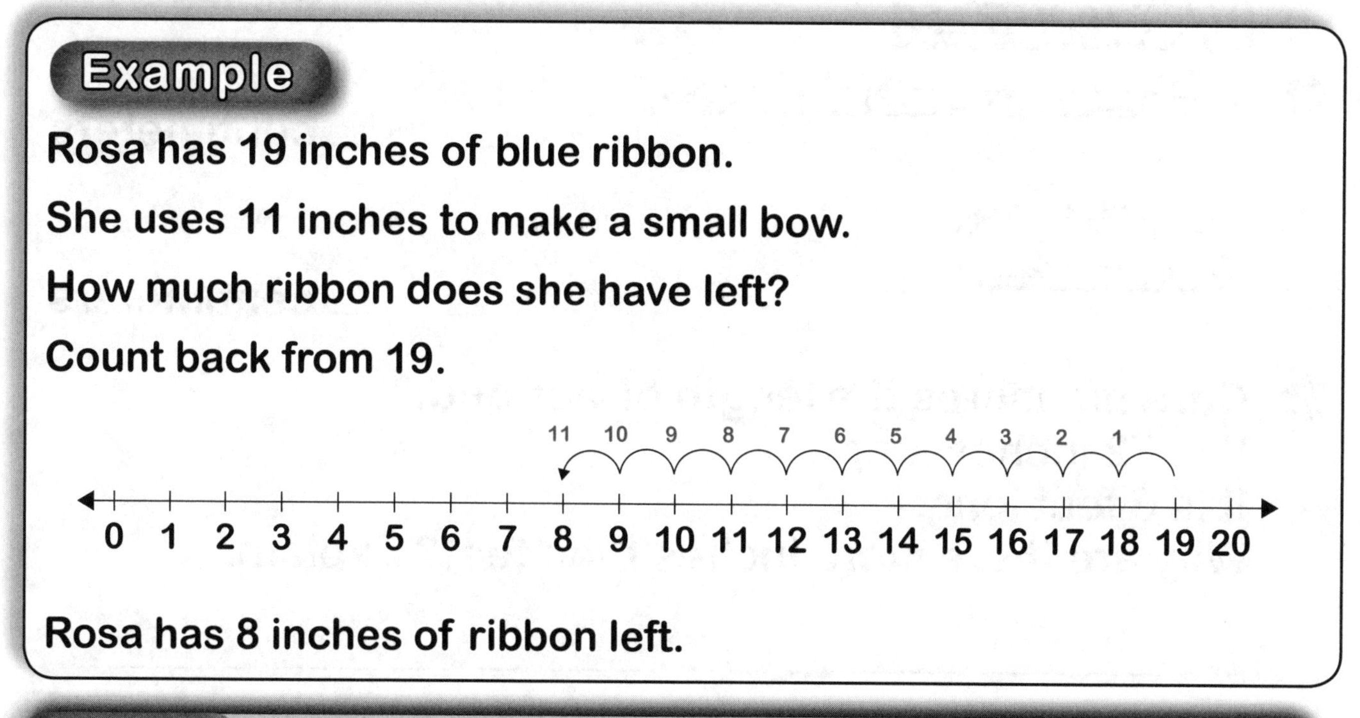

Solve

Use the number line to solve.

1. The roll of paper had 20 feet of paper. Ron used 3 feet for an art project. How many feet are left?

   17 feet

2. Jess painted a wall that is 12 feet wide. Dean painted a wall that is 10 feet wide. How much wider was the wall Jess painted?

   ______ feet wider

Name ____________________

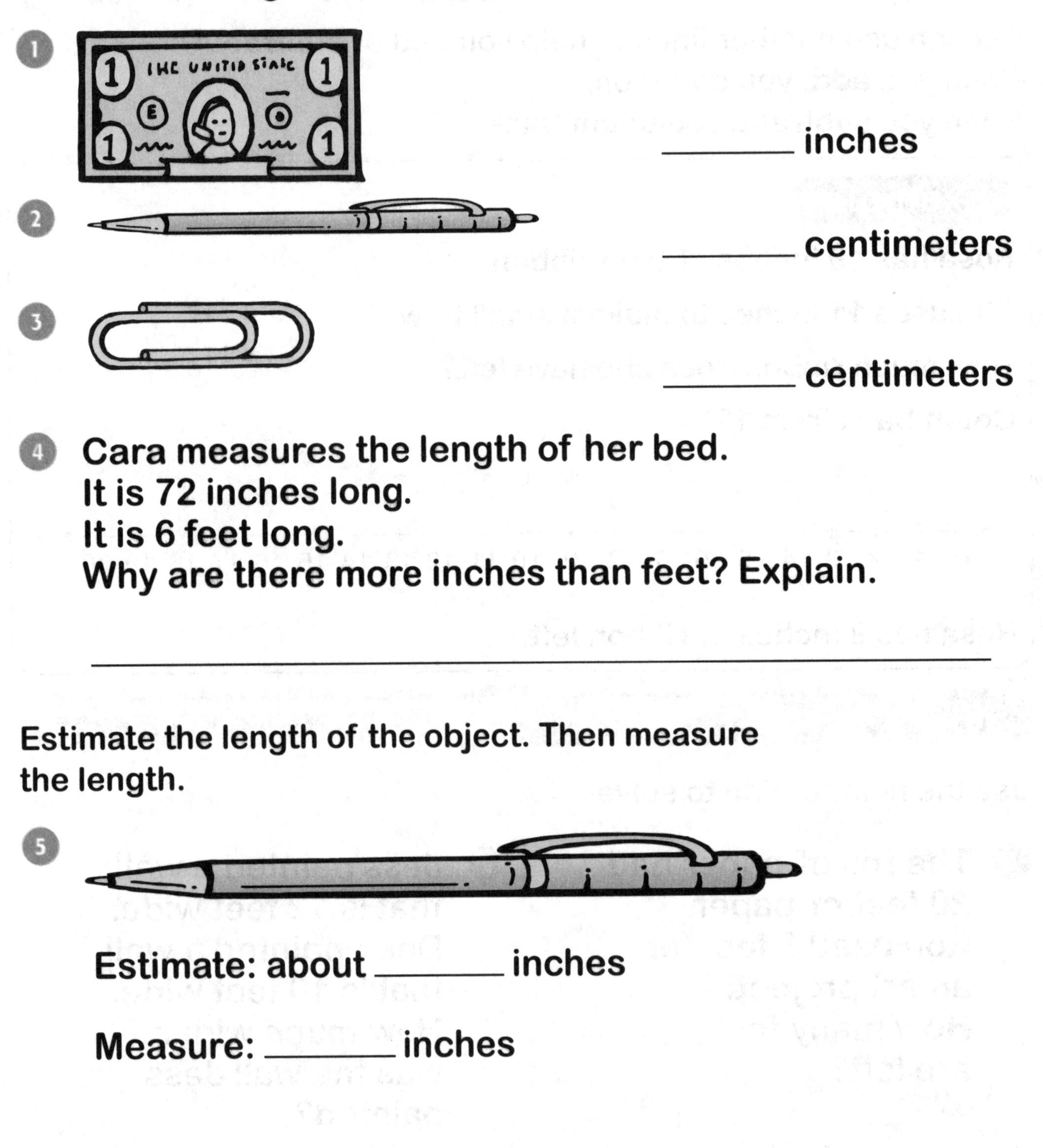

Measure the length of each object.

1 ______ inches

2 ______ centimeters

3 ______ centimeters

4 Cara measures the length of her bed.
It is 72 inches long.
It is 6 feet long.
Why are there more inches than feet? Explain.

____

Estimate the length of the object. Then measure the length.

5

Estimate: about ______ inches

Measure: ______ inches

Name ______________________

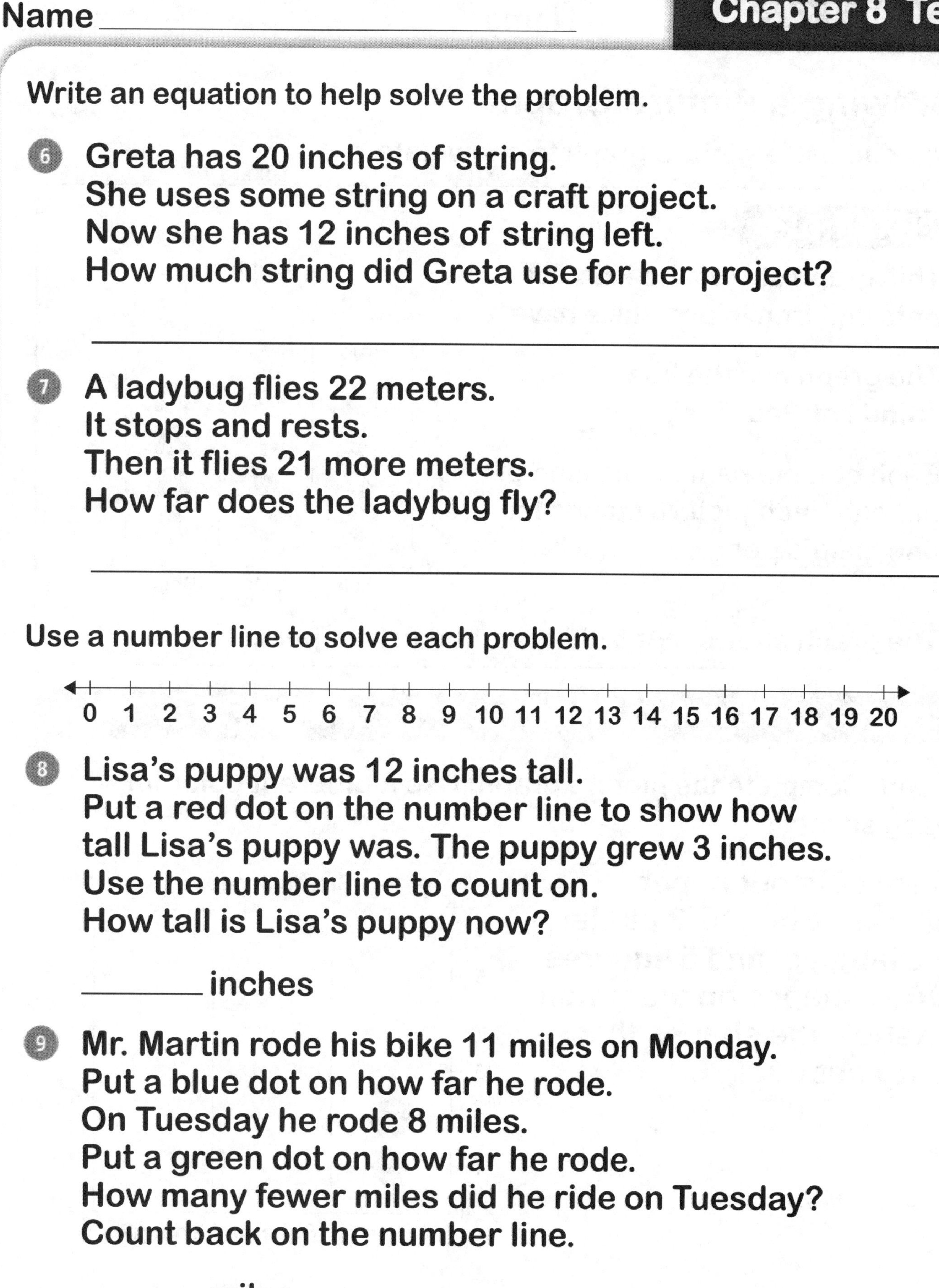

Write an equation to help solve the problem.

6. Greta has 20 inches of string.
She uses some string on a craft project.
Now she has 12 inches of string left.
How much string did Greta use for her project?

____

7. A ladybug flies 22 meters.
It stops and rests.
Then it flies 21 more meters.
How far does the ladybug fly?

____

Use a number line to solve each problem.

8. Lisa's puppy was 12 inches tall.
Put a red dot on the number line to show how tall Lisa's puppy was. The puppy grew 3 inches.
Use the number line to count on.
How tall is Lisa's puppy now?

_______ inches

9. Mr. Martin rode his bike 11 miles on Monday.
Put a blue dot on how far he rode.
On Tuesday he rode 8 miles.
Put a green dot on how far he rode.
How many fewer miles did he ride on Tuesday?
Count back on the number line.

_______ miles

Name ______________________

Drawing a Picture Graph

You can use a picture graph to show data.

Example

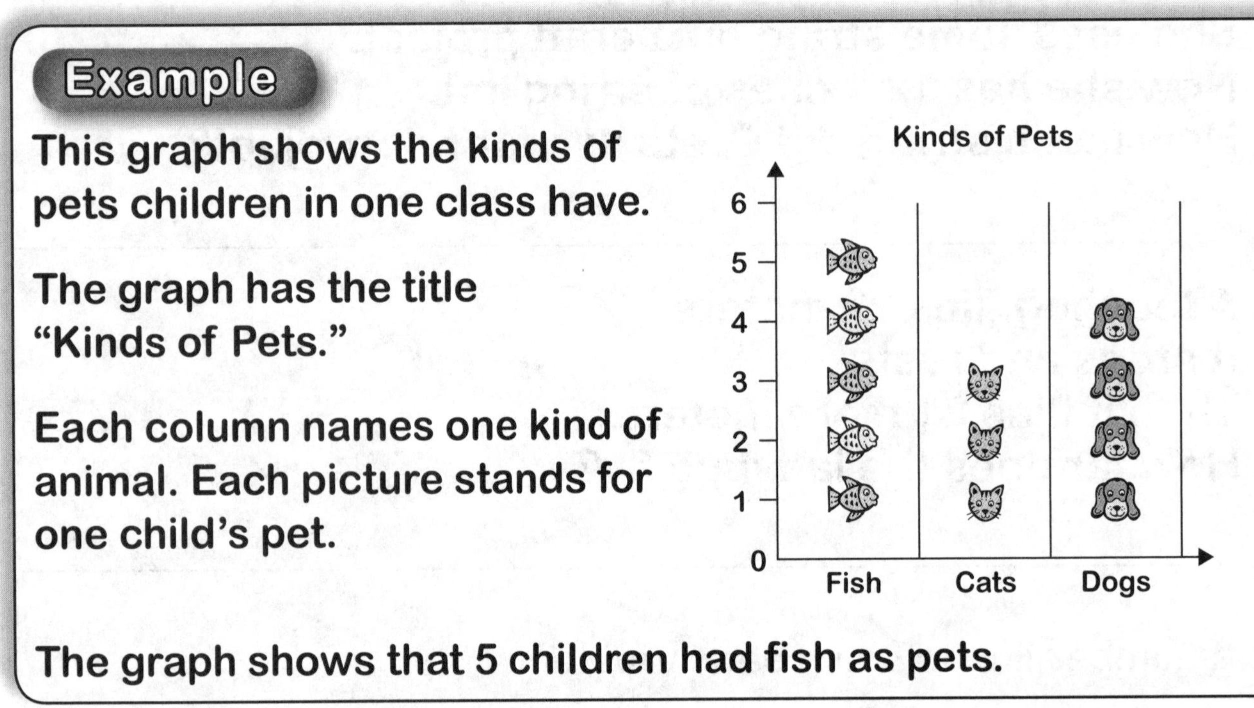

This graph shows the kinds of pets children in one class have.

The graph has the title "Kinds of Pets."

Each column names one kind of animal. Each picture stands for one child's pet.

The graph shows that 5 children had fish as pets.

Draw

Read. Complete the picture graph. Use a different color for each shape.

Suzy cuts out paper shapes. She has 2 circles, 3 triangles, and 5 squares. Draw shapes on the graph to show the shapes that Suzy cut out.

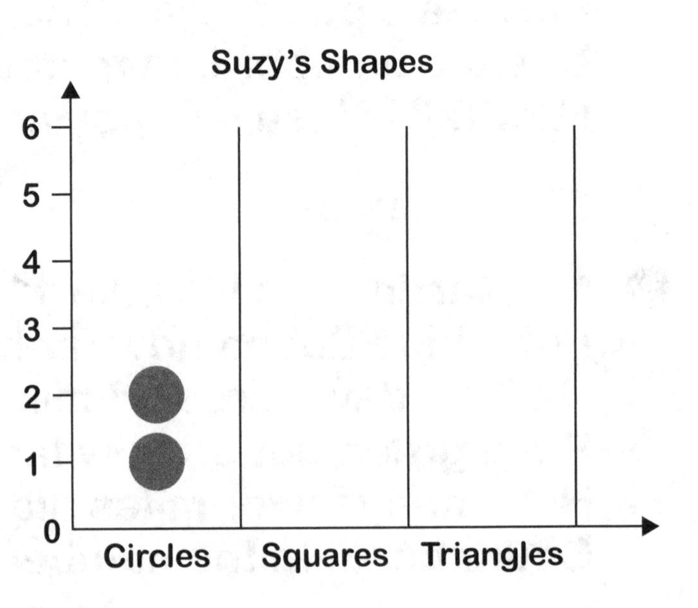

Name ______________________________

Drawing a Bar Graph

You can use a bar graph to show data.

Example

This graph shows how old four cousins are. The names are below the bars showing the ages.

The graph shows that the oldest child is Christina.

Two children are the same age.

The youngest child is 3 years old.

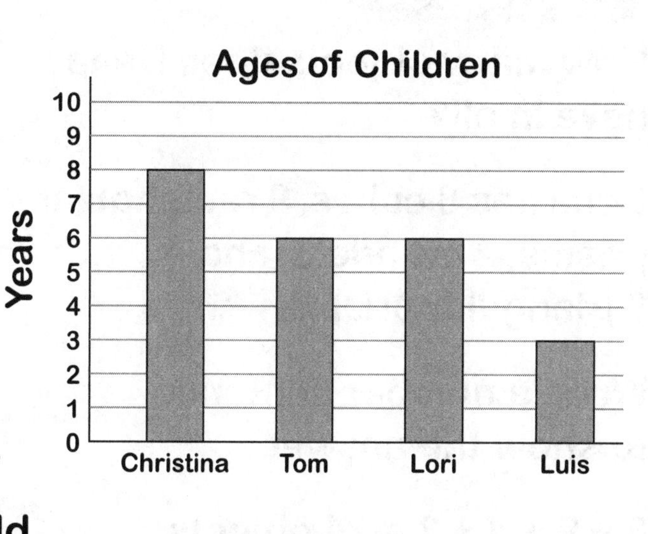

Draw

Read. Complete the bar graph.

In the vase are red, yellow, and blue flowers.

There are 7 blue flowers, 3 red flowers, and 6 yellow flowers. Draw bars on the graph to show the numbers of flowers in the vase.

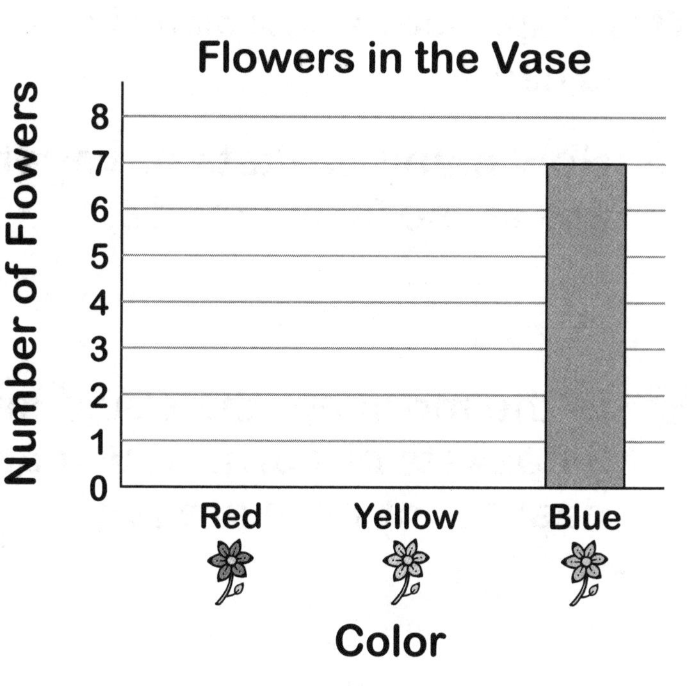

Name ____________________

Using a Bar Graph to Solve Problems

You can use a bar graph to solve problems.

Example

How many objects does Dena have in all?

Dena has 8 cubes, 9 rectangular prisms, 4 cylinders, and 3 triangular prisms.

Write a number sentence to show the answer.

$8 + 9 + 4 + 3 = 24$ objects

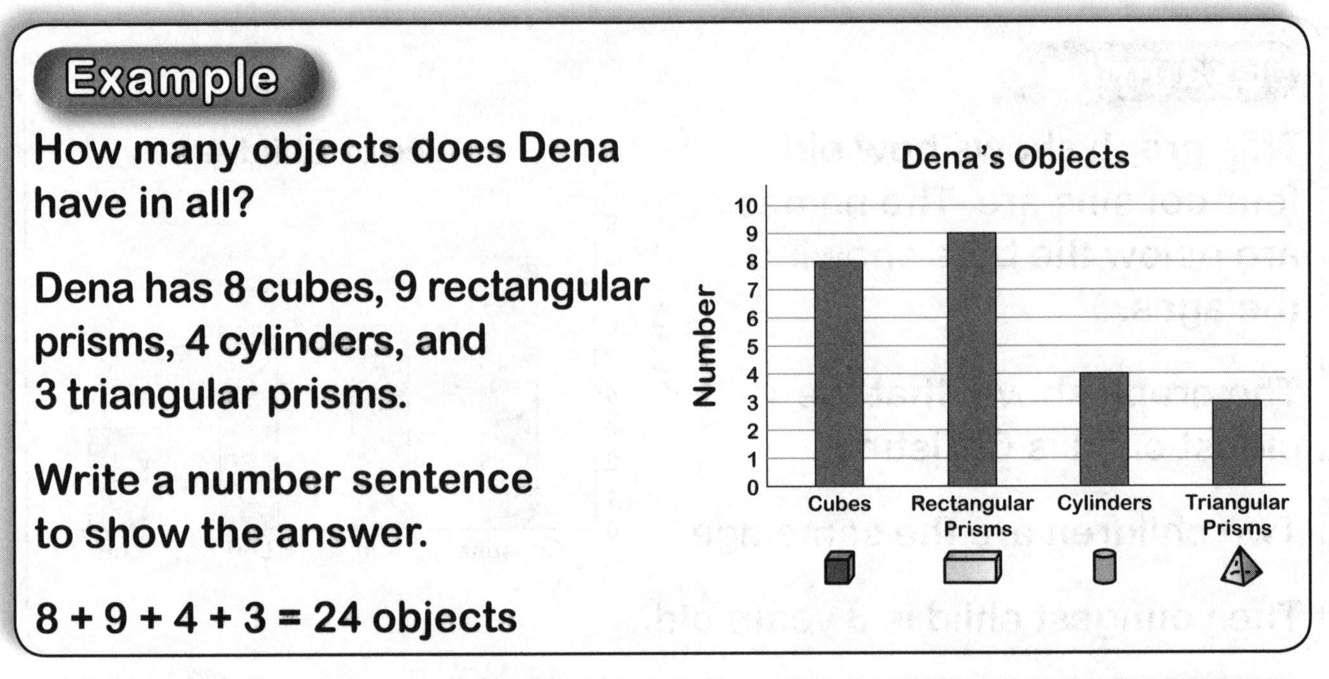

Solve

Use the graph to help you answer the questions.
This graph shows how many baskets of vegetables a farm stand had.

1. How many baskets of vegetables are at the farm stand?

   ____

2. In the morning, the stand sells 3 baskets of corn. How many baskets of corn are left?

   ____

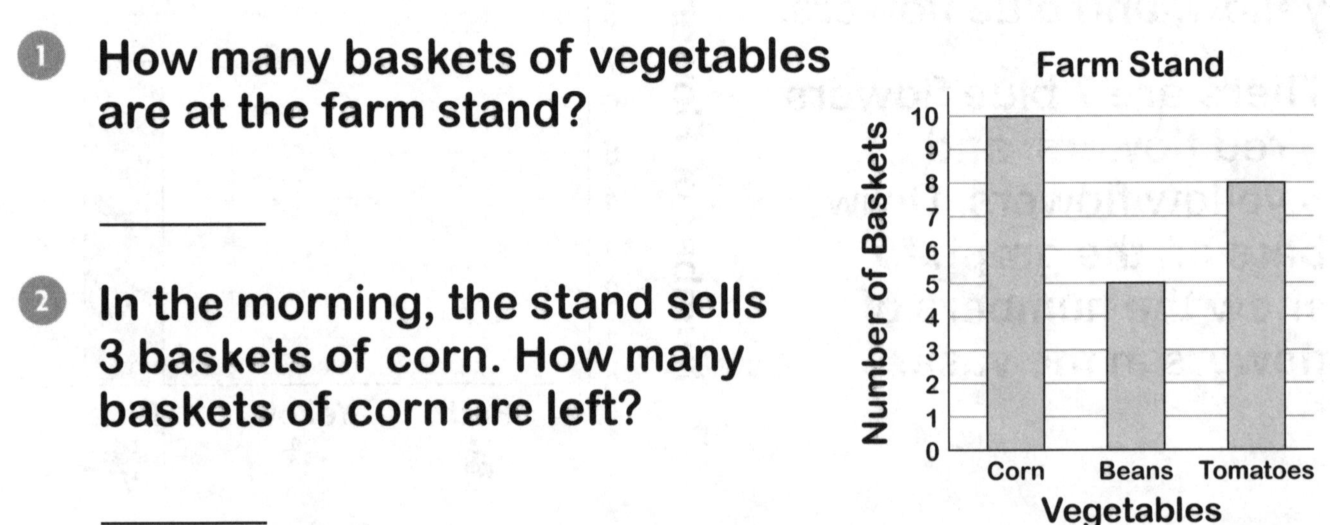

Name ____________________

Using a Bar Graph to Compare Data

You can use a bar graph to show and compare data.

Example

Which supply does Emma have most of?

Look at each bar. The bar for Pencils is the longest.

There are more pencils than any other supply.

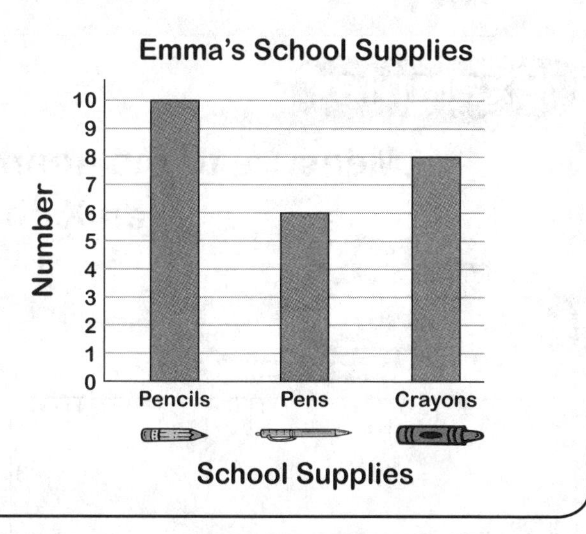

Solve

Use the graph to help you answer the questions.
This graph shows the fruit children in one class like best.

1. How many more children liked bananas than liked apples?

   ____

2. How many children did not choose oranges as their favorite fruit?

   ____

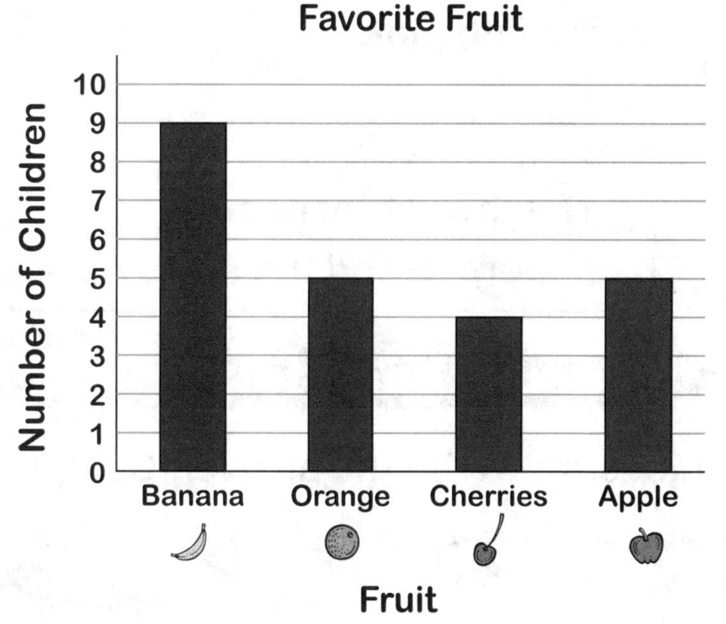

Name ______________________

Making a Line Plot to Show Measurements

You can measure objects and show their lengths on a line plot. Then you can see which are the same length.

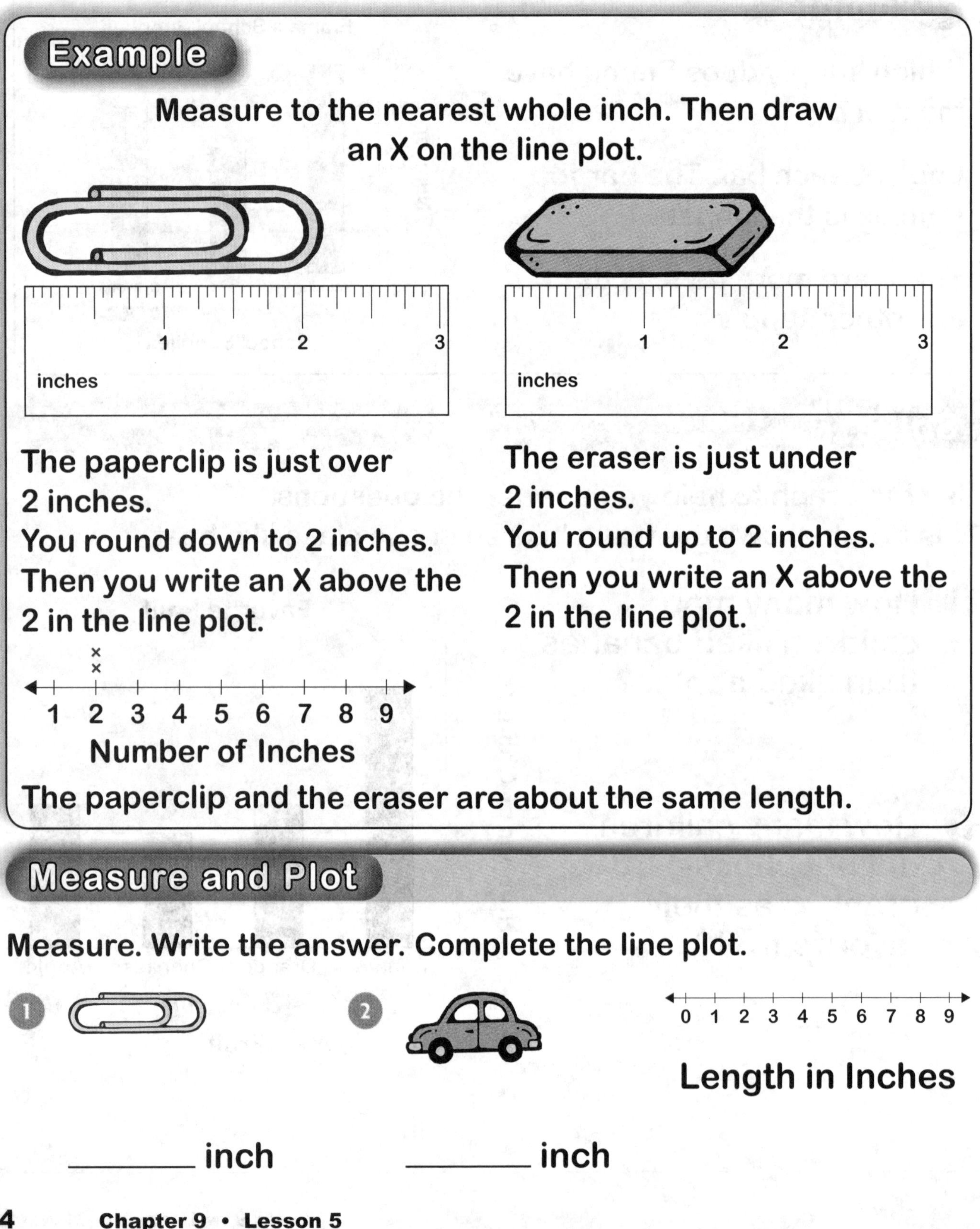

Example

Measure to the nearest whole inch. Then draw an X on the line plot.

The paperclip is just over 2 inches.
You round down to 2 inches.
Then you write an X above the 2 in the line plot.

The eraser is just under 2 inches.
You round up to 2 inches.
Then you write an X above the 2 in the line plot.

Number of Inches

The paperclip and the eraser are about the same length.

Measure and Plot

Measure. Write the answer. Complete the line plot.

1. ________ inch

2. ________ inch

Length in Inches

Name ______________________

Measuring Two Times and Making a Line Plot

Measure the same object two times.
That way you are sure you have measured correctly.

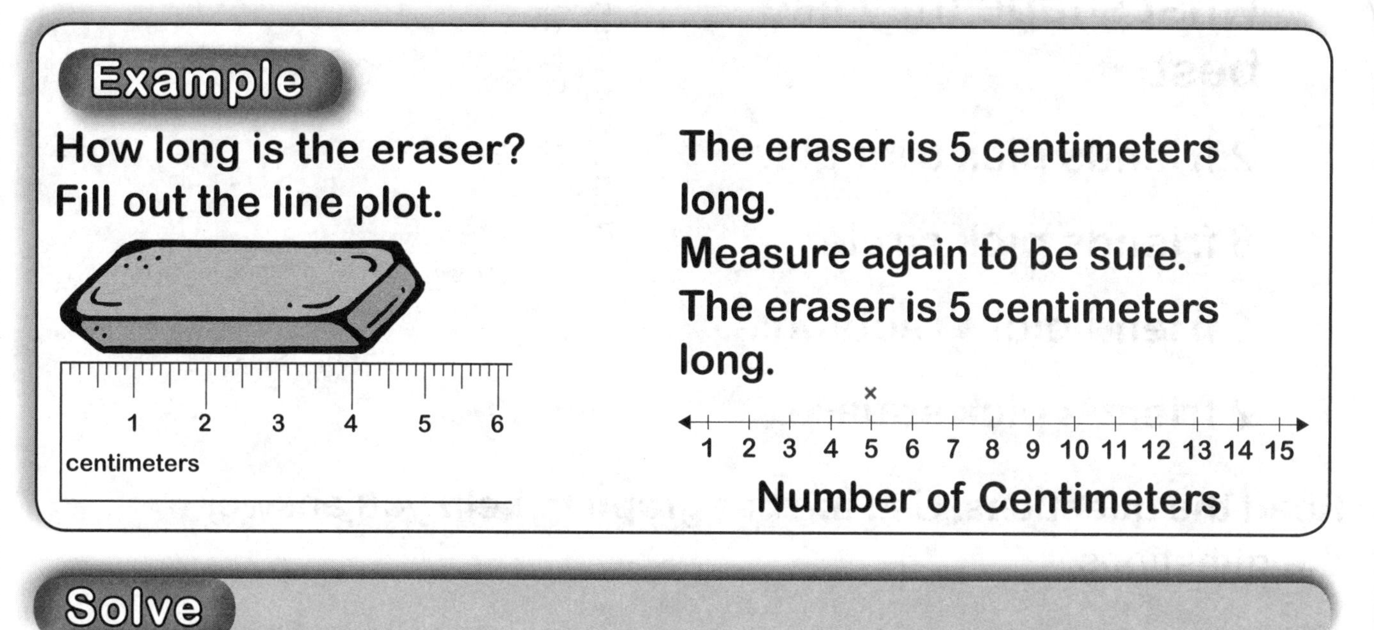

Solve

Use a metric ruler. Measure each item two times. Write the answer. Then put an X on the line plot.

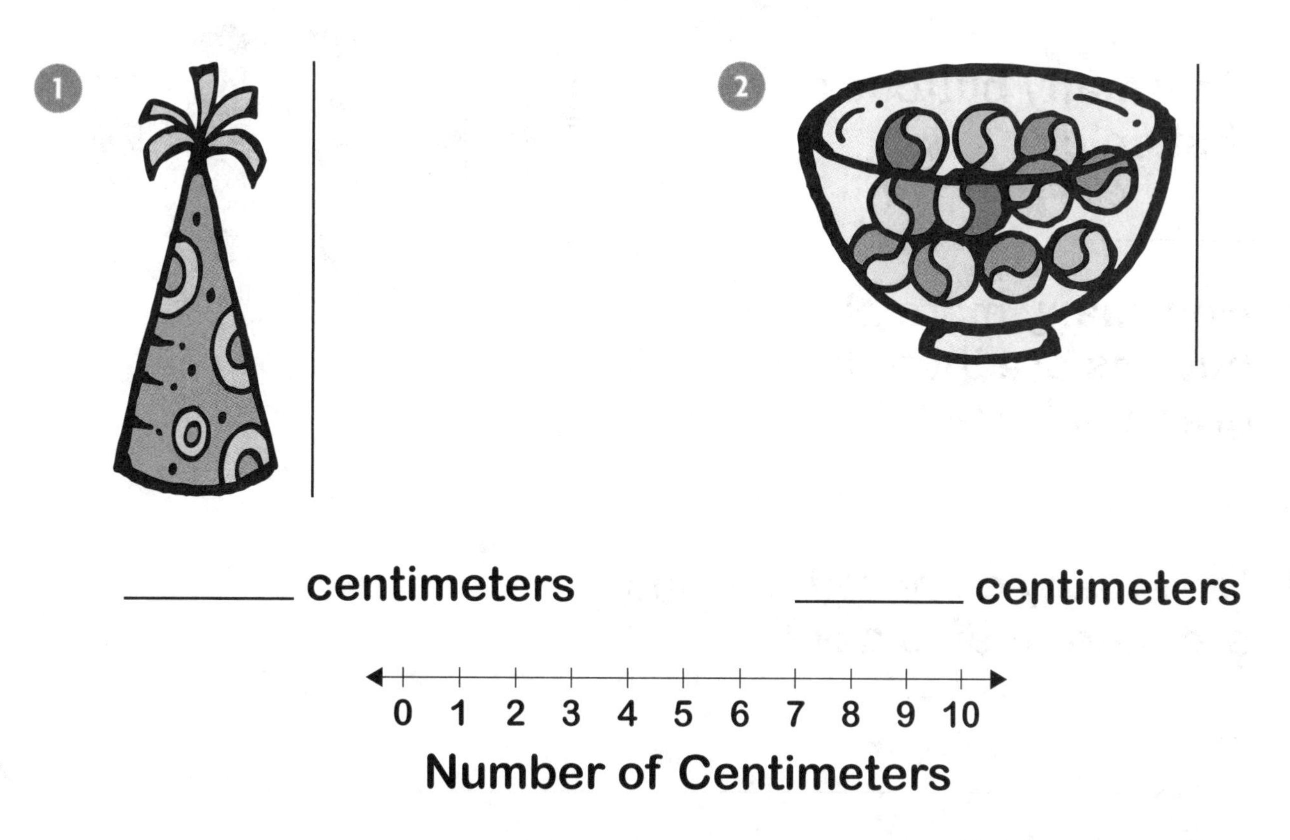

Chapter 9 Test

Name ____________________

Read. Complete the bar graph.

1. Lin asks her friends what shape they like best.

   2 friends pick triangle.

   6 friends pick circle.

   1 friend picks rectangle.

   2 friends pick square.

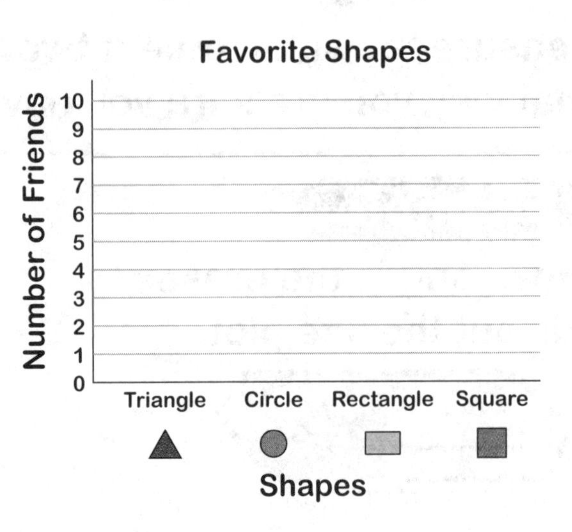

Read the questions. Use the bar graph to help you answer the questions.

2. Which color has the most buttons?

   ____

   How many buttons are that color?

   ____

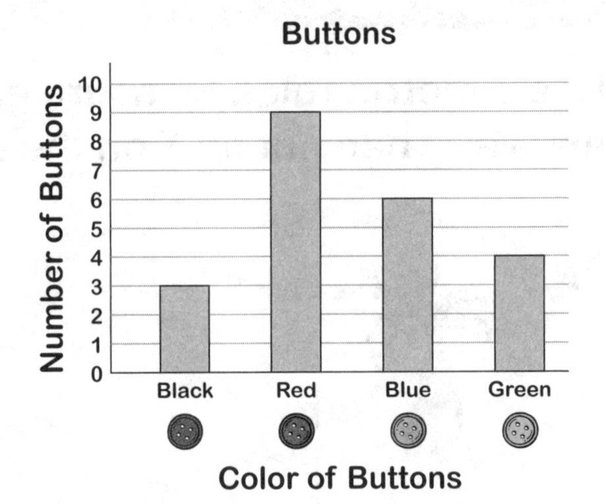

3. How many more blue buttons are there than green buttons?

   ____

4. How many more red buttons are there than black?

   ____

Name ______________________________

Use a ruler. Measure to the nearest centimeter. Write the answer.

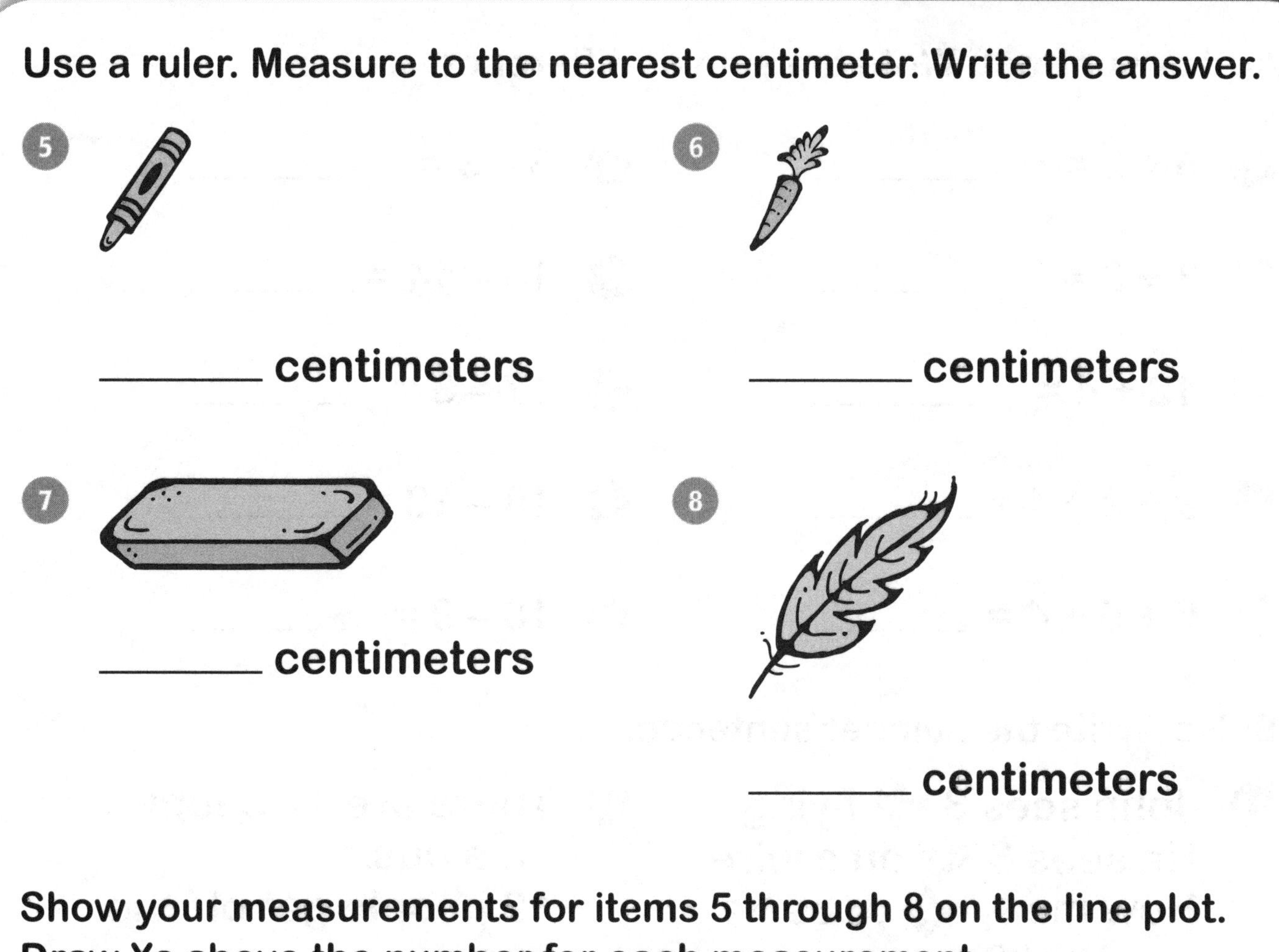

5 ________ centimeters

6 ________ centimeters

7 ________ centimeters

8 ________ centimeters

Show your measurements for items 5 through 8 on the line plot. Draw Xs above the number for each measurement.

9

0 1 2 3 4 5 6 7 8 9 10

Number of Centimeters

Look at your line plot. Answer the question.

10 Which length has two items?

________ centimeters

Chapters 1-9 Review

Name ____________________

Add or subtract. Write the sum or difference.

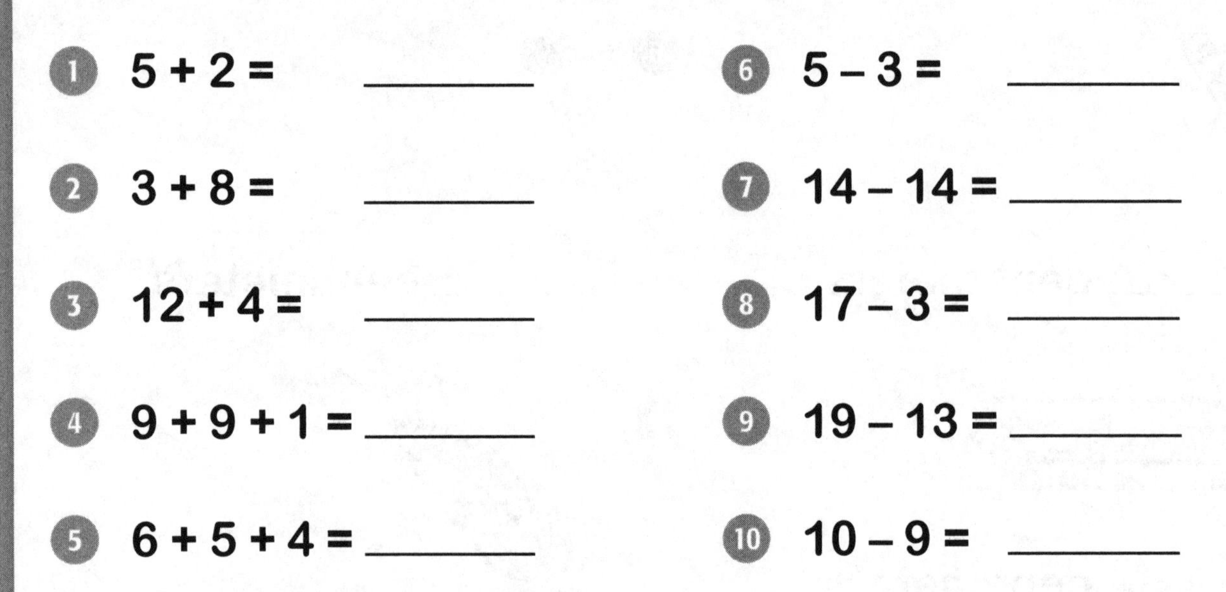

1. 5 + 2 = ______
2. 3 + 8 = ______
3. 12 + 4 = ______
4. 9 + 9 + 1 = ______
5. 6 + 5 + 4 = ______
6. 5 – 3 = ______
7. 14 – 14 = ______
8. 17 – 3 = ______
9. 19 – 13 = ______
10. 10 – 9 = ______

Solve. Write the number sentence.

11. John sees 3 birds flying.
He sees 6 birds on a wire.
How many birds does he see in all?

____ + ____ = ____ birds

12. Alice has 13 stamps.
She gives 8 to Tammy.
Then Alice buys 5 more stamps.
How many stamps does Alice have now?

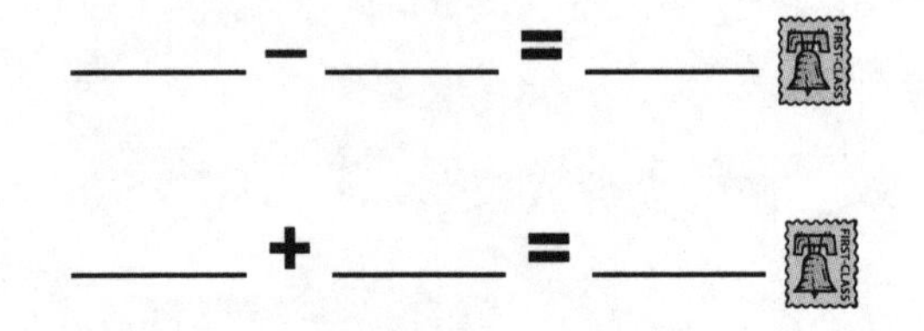

____ – ____ = ____ stamps

____ + ____ = ____ stamps

13. There are 17 people on a bus.
12 people get off the bus.
How many people are on the bus now?

____ – ____ = ____ people

14. Ms. Lee has 13 grapes.
She eats 6 grapes.
How many grapes are left?

____ – ____ = ____ grapes

Name ____________________

Add or subtract. Write the sum or difference.

(15) $41 + 25$ = ____

(16) $18 + 45$ = ____

(17) $67 - 31$ = ____

(18) $42 - 28$ = ____

(19) $83 + 15$ = ____

(20) $44 + 14$ = ____

(21) $78 - 78$ = ____

(22) $91 - 55$ = ____

Add or subtract to solve. Show your work.

(23) Mr. Brown plants
26 flowers by the shed.
He plants 45 flowers
by the house.
How many flowers does
Mr. Brown plant?

____ flowers

(24) There are 53 muffins
in the bake shop.
The shop sells
28 muffins.
How many muffins
are left?

____ muffins

Solve for the unknown number. Ask yourself:
Do I need to add or subtract to find the unknown number?
Write the answer on the line.

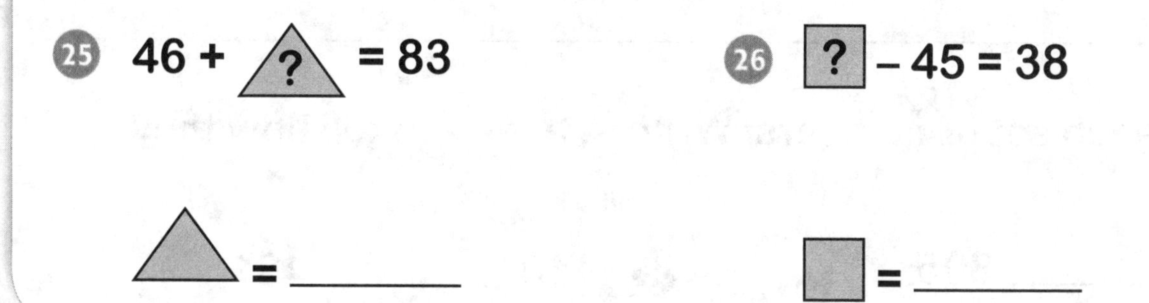

Name ____________________

Tell whether a number is odd or even.
Use paper clips or other small objects.
Try to put the objects in pairs. Then circle **odd** or **even**.

27. 6 odd even

28. 17 odd even

29. 12 odd even

30. 9 odd even

Write a number sentence to tell what the array shows.

31. ____ + ____ + ____ = ____

32. ____ + ____ = ____

33. Write the number in words.

____

34. Write the number in expanded form.

____

Compare each set of numbers. Write <, >, or = to tell how they compare.

35. 539 ________ 395

36. 413 ________ 431

Name ______________________

Chapters 1-9 Review

Add or subtract. Write the sum or difference.

37 728 − 502	**38** 673 − 295	**39** 635 − 100	**40** 581 − 100
41 128 + 802	**42** 327 + 184	**43** 220 314 + 372	**44** 470 414 + 72

Solve. Write your answer.

45 Judy sees this problem in a book: 220 + ? = 538. How can she find the missing addend?

____

____

Look at each clock. Write the time with A.M. or P.M.

Afternoon

46

____

Morning

47

____

Count the money. Write the total. Use the ¢ or $ symbol.

48

____

 Name ____________________

49 Draw a shape with 3 angles.

50 Which shape is a cube? Circle it.

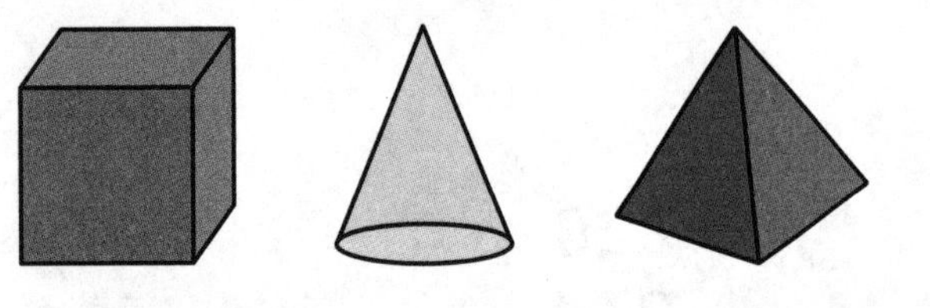

51 Draw a box around the circle that shows thirds.

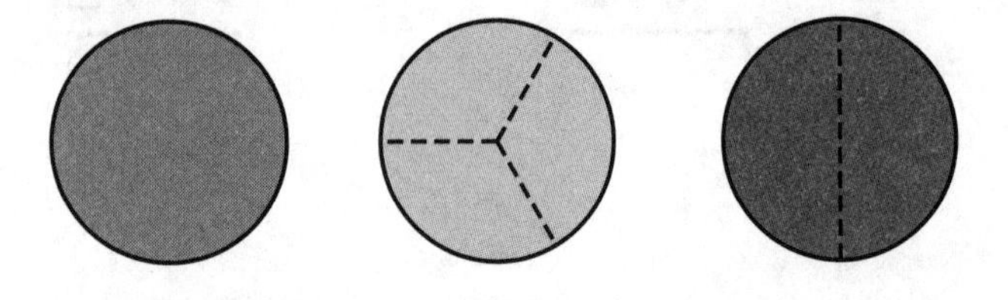

52 Which shapes have half colored blue? Circle them.

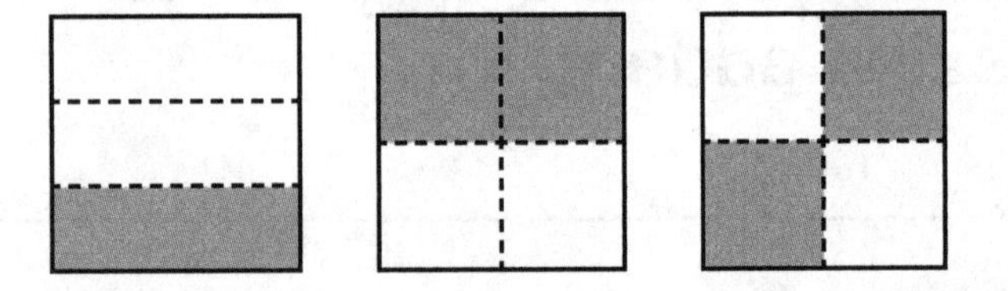

53 Sharon measures the length of a bike 2 times.
She measures the bike in feet. It is 6 feet long.
Then she measures the bike in yards. It is 2 yards long.
Why are there more feet than yards?

____

____

54 Find a water glass at home. Estimate how tall it is.
Then measure the glass.

Estimate: about ________ centimeters

Measure: ________ centimeters

Name ______________________

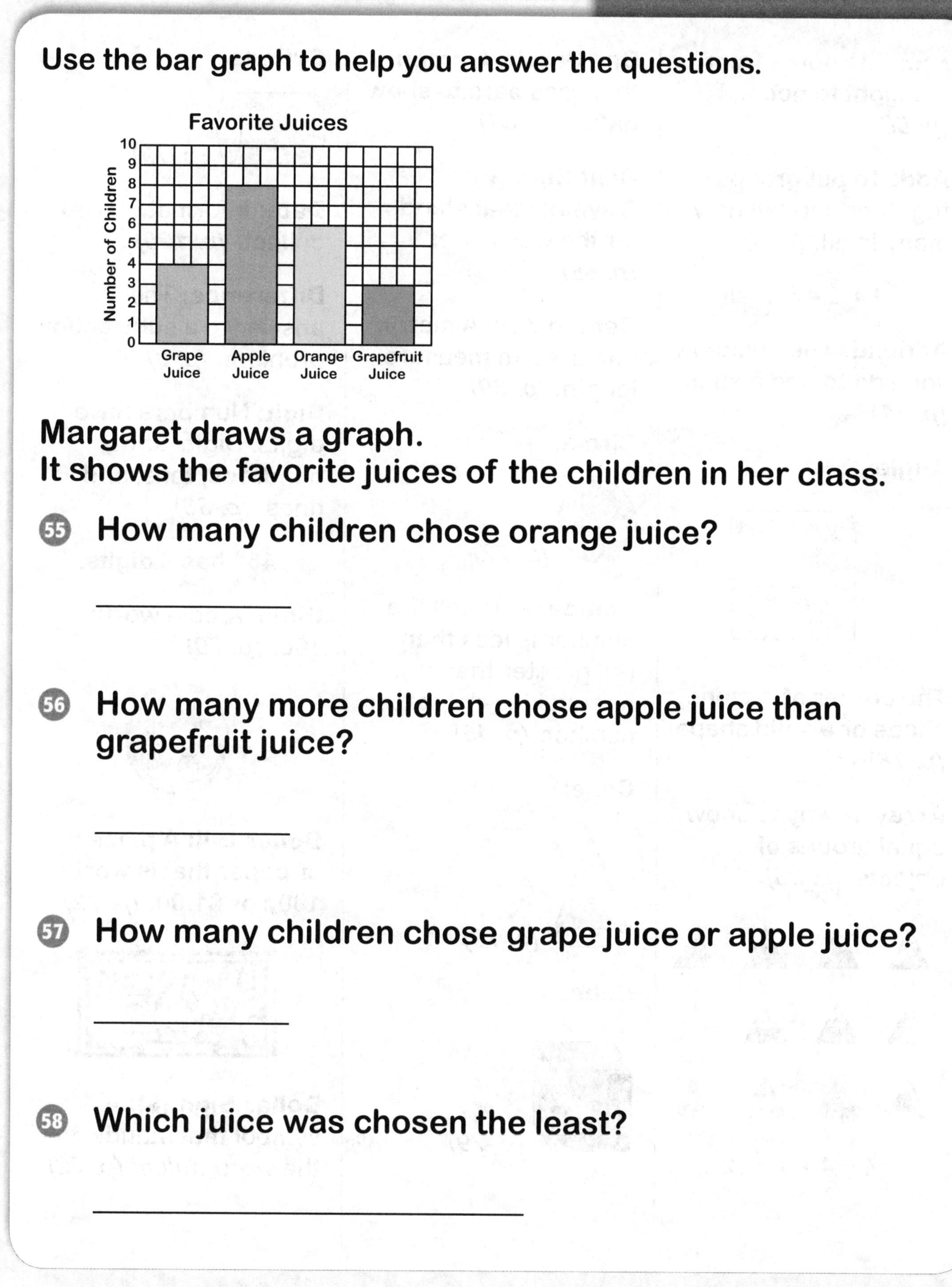

Use the bar graph to help you answer the questions.

Margaret draws a graph.
It shows the favorite juices of the children in her class.

55. How many children chose orange juice?

____

56. How many more children chose apple juice than grapefruit juice?

____

57. How many children chose grape juice or apple juice?

____

58. Which juice was chosen the least?

____

Picture Dictionary

A.M.: The hours from midnight to noon. *(p. 66)*

Add: To put groups together and tell how many in all. *(p. 8)*

1 + 2 = 3 in all

Addend: The numbers you add to find a sum. *(p. 11)*

Angle:

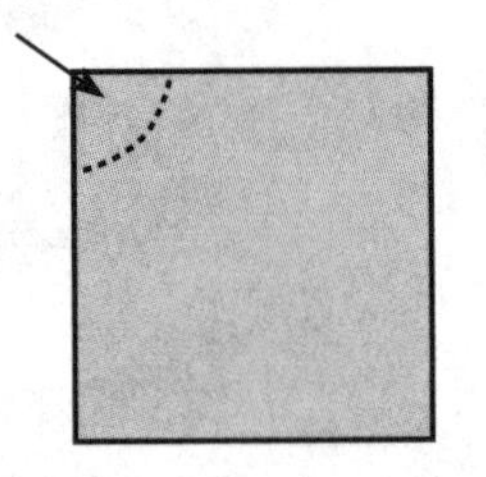

The corner of a plane shape or a solid shape. *(p. 78)*

Array: A way to show equal groups of objects. *(p. 33)*

4 + 4 + 4 = 12

Bar Graph: A graph that uses bars to show data. *(p. 101)*

Cent Sign (¢): A symbol that stands for the word *cent.* *(p. 68)*

Centimeter: A metric unit used to measure length. *(p. 89)*

Circle:

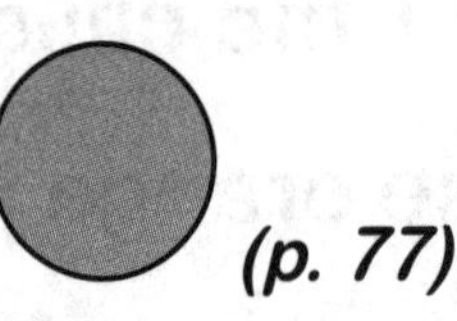

(p. 77)

Compare: To tell if a number is less than (<), greater than (>), or equal to (=) another number. *(p. 45)*

Cone:

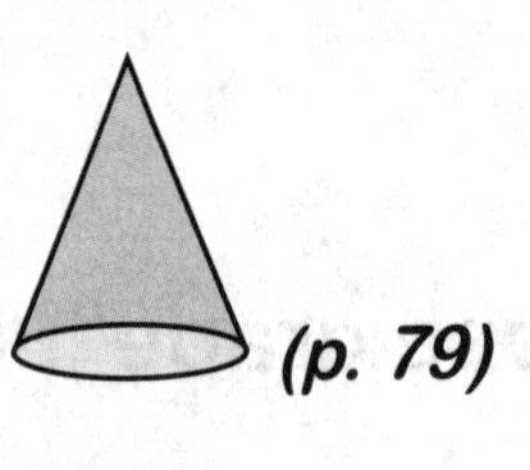

(p. 79)

Cube:

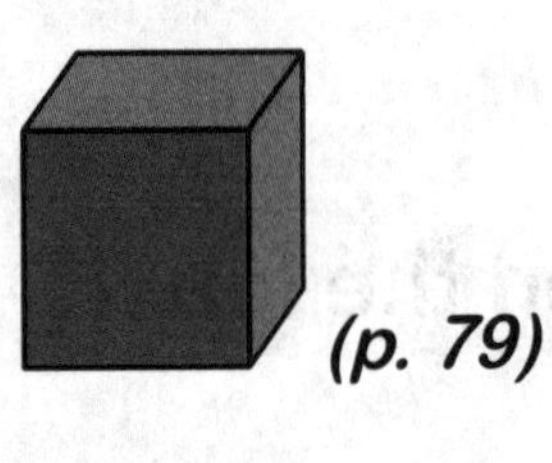

(p. 79)

Cylinder:

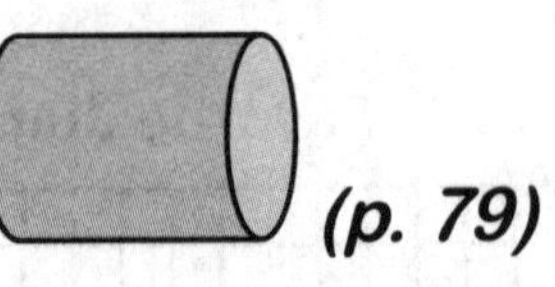

(p. 79)

Data: Information you collect. *(p. 100)*

Difference: The answer to a subtraction problem. *(p. 9)*

Digit: Numbers have digits. Digits show hundreds, tens, and ones. *(p. 38)*

461 has 3 digits.

Dime: A coin worth 10¢. *(p. 70)*

Dollar Bill: A piece of paper that is worth 100¢ or $1.00. *(p. 72)*

Dollar Sign ($): A symbol that stands for the word *dollar.* *(p. 72)*

Equal Parts: Parts that are the same number and size. *(p. 81)*

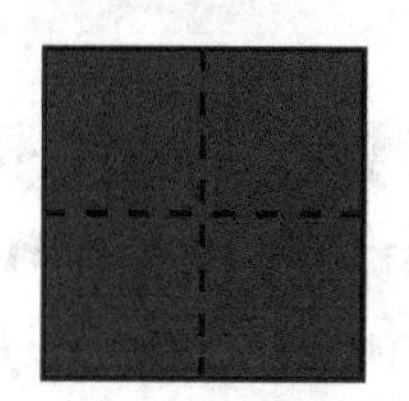

Equal To (=): To have the same value or amount. *(p. 45)*

3 = 2 + 1

Estimate: A careful guess. *(p. 92)*

Even Number: A number that ends with 0, 2, 4, 6, or 8. *(p. 29)*

Expanded Form: Numbers that show the values of each digit. *(p. 44)*

129 = 100 + 20 + 9

Foot, Feet: 12 inches. *(p. 88)*

Fourth, Fourths: Four equal parts of a whole. *(p. 81)*

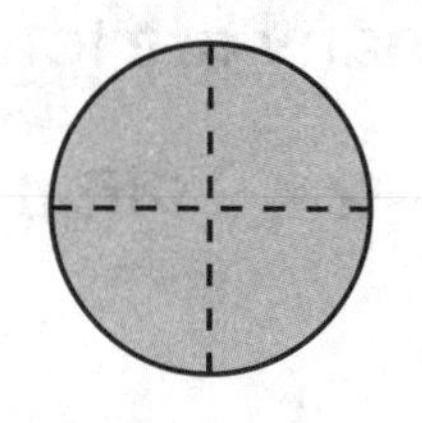

Greater Than (>): *(p. 45)*

3 > 2

3 is greater than 2.

Half, Halves: Two equal parts of a whole. *(p. 81)*

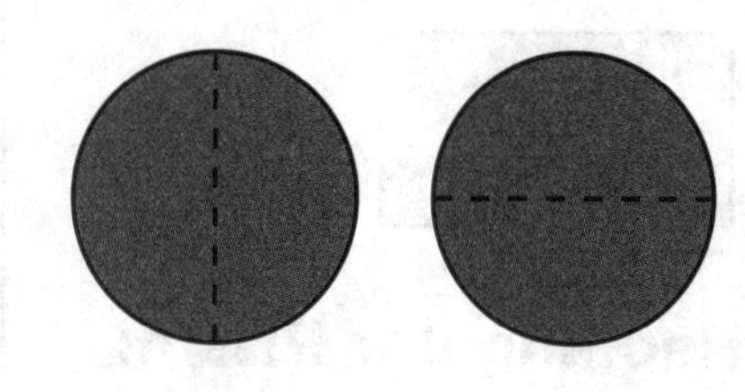

Hexagon:

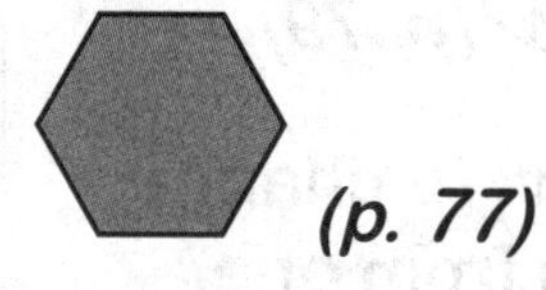

(p. 77)

Inch: A unit used to measure length. *(p. 88)*

Length: The distance from one end of an object to another. *(p. 88)*

Less Than (<): *(p. 45)*

3 < 4

3 is less than 4.

Line Plot: A way to show data on a line. *(p. 104)*

Measure: To find how much there is of something. *(p. 88)*

Measuring Tape: A tool used to measure length. *(p. 88)*

Mental Math: Math you do in your head. *(p. 60)*

$$\begin{array}{r} 62 \\ +\ 10 \\ \hline 72 \end{array}$$

Meter: 100 centimeters. *(p. 89)*

Nickel: A coin worth 5¢. *(p. 69)*

Number Name: Words that tell numbers. *(p. 43)*

fifteen 15

Odd Number: A number that ends with 1, 3, 5, 7, or 9. *(p. 29)*

Picture Dictionary

P.M.: The hours from noon to midnight. *(p. 66)*

Penny: A coin worth 1¢. *(p. 68)*

Pentagon:

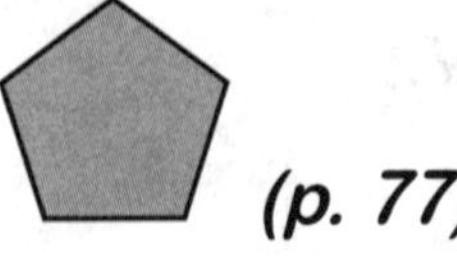

(p. 77)

Picture Graph: A graph that uses pictures to show data. *(p. 100)*

Plane Shape: A flat shape, like a circle or a square. *(p. 77)*

Pyramid:

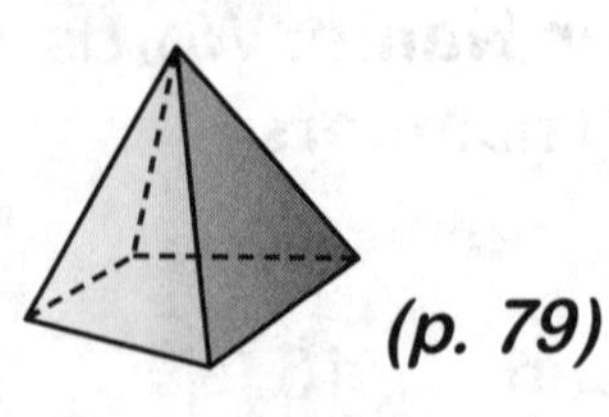

(p. 79)

Quadrilateral: Any plane shape with 4 sides. *(p. 77)*

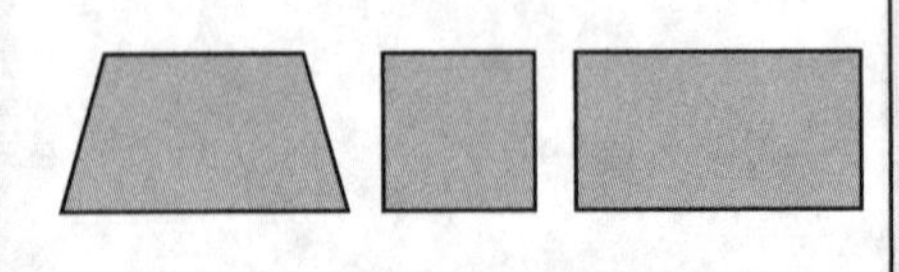

Quarter: A coin worth 25¢. *(p. 72)*

Rectangle:

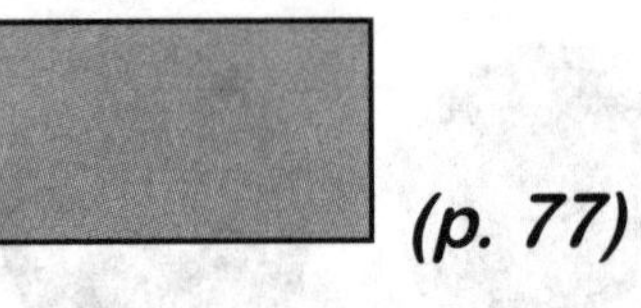

(p. 77)

Rectangular Prism:

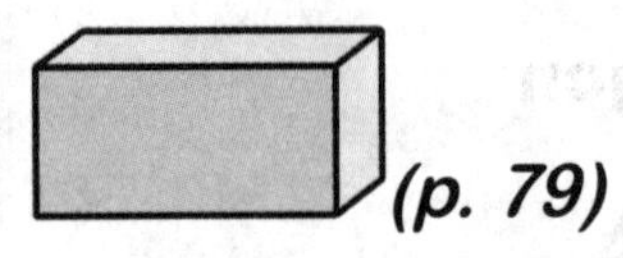

(p. 79)

Regroup: To change numbers from one value to another. *(p. 19)*

Regroup 10 ones as 1 ten.

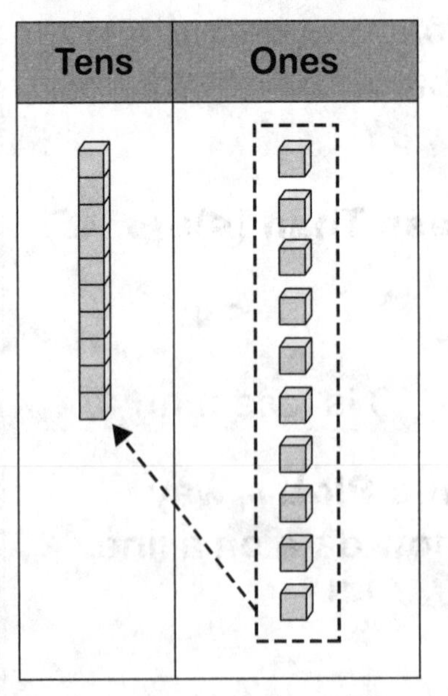

Ruler: A tool used to measure length. *(p. 88)*

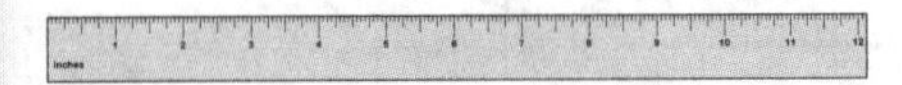

Skip Count: To count by something other than ones. *(p. 39)*

Skip count by 5s.

5, 10, 15, 20, ...

Solid Shape: A shape that is not flat. *(p. 79)*

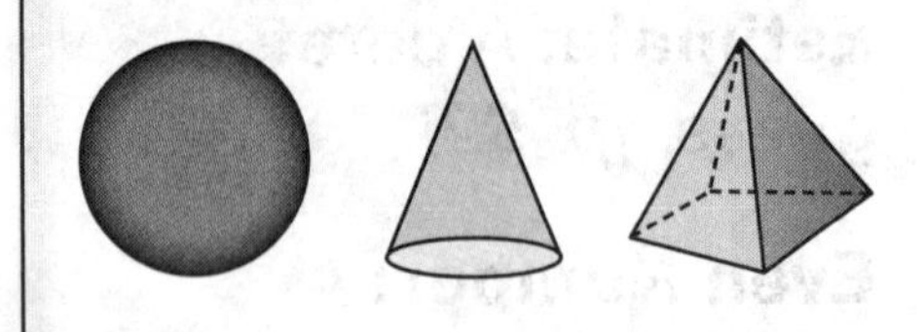

Sphere:

(p. 79)

Square:

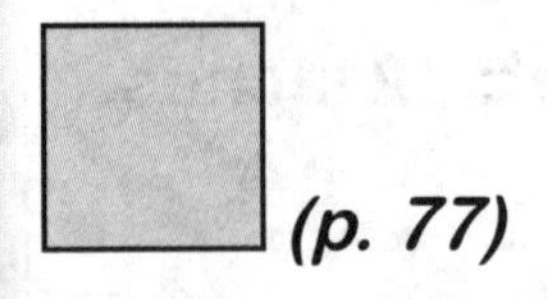

(p. 77)

Subtract: To take groups away and tell how many are left. *(p. 9)*

4 – 2 = 2 left over

Sum: The answer to an addition problem. *(p. 8)*

Symbol: A picture that stands for something else. *(p. 26)*

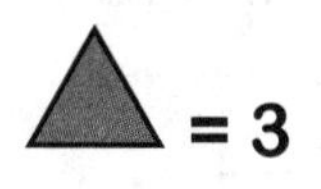

Third, Thirds: Three equal parts of a whole. *(p. 81)*

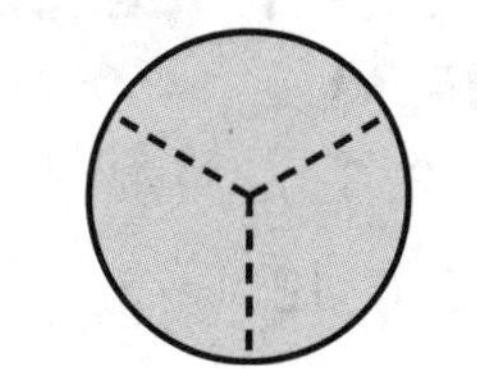

Triangle:

(p. 77)

Yard: 3 feet. *(p. 88)*

Yardstick: A tool used to measure length. *(p. 88)*

Chapter 1

Chapter 1 • Lesson 1 Page 8

1. 8
2. 12
3. 2
4. 20
5. 17
6. 18

Chapter 1 • Lesson 2 Page 9

1. 15
2. 8
3. 3
4. 1
5. 11
6. 10
7. 2
8. 0
9. 12
10. 8
11. 2
12. 3
13. 8
14. 6

Chapter 1 • Lesson 3 Page 10

1. 20, 7
2. 18, 9
3. 14, 3
4. 16, 2
5. 9, 3
6. 10, 5

Chapter 1 • Lesson 4 Page 11

1. 4, 2, 6
2. 12, 6, 18
3. 3, 3, 6

Chapter 1 • Lesson 5 Page 12

1. 7, 3, 4
2. 17, 5, 12

Chapter 1 • Lesson 6 Page 13

1. 5, 2, 3;
   3, 4, 7
2. 6, 4, 2;
   2, 6, 8

Chapter 1 • Lesson 7 Page 14

1. 7, 6, 13;
   13, 9, 4
2. 11, 5, 16;
   16, 8, 8

Chapter 1 • Lesson 8 Page 15

Drawings for items 1–2 will vary.

1. 8 – 3 = 5
2. 5 + 9 = 14

Chapter 1 Test Pages 16-17

1. 6
2. 17
3. 19
4. 12
5. 13
6. 1
7. 0
8. 14
9. 6 + 4 = 10
10. 5 + 6 + 8 = 19
11. 8 + 3 = 11;
    11 – 7 = 4
12. 5 – 4 = 1;
    1 + 7 = 8
13. 11 + 7 = 18

Chapter 2

Chapter 2 • Lesson 1 Page 18

1. 58
2. 98
3. 66
4. 68
5. 82
6. 49
7. 77
8. 99
9. 67

Chapter 2 • Lesson 2 Page 19

1. 25
2. 80
3. 65

4. 77
5. 91
6. 84

Chapter 2 • Lesson 3
Page 20

1. 23
2. 12
3. 51
4. 10
5. 11
6. 17
7. 12
8. 42
9. 53

Chapter 2 • Lesson 4
Page 21

1. 68
2. 8
3. 15
4. 29
5. 16
6. 57

Chapter 2 • Lesson 5
Page 22

1. 55 + 32 = 87
2. 12 + 9 = 21

Chapter 2 • Lesson 6
Page 23

1. 37 – 25 = 12
2. 77 – 49 = 28

Chapter 2 • Lesson 7
Page 24

1. 38 – 12 = 26;
   26 + 25 = 51
2. 71 – 46 = 25;
   25 + 20 = 45

Chapter 2 • Lesson 8
Page 25

1. 35 + 52 = 87;
   87 – 46 = 41
2. 39 + 43 = 82;
   82 – 58 = 24

Chapter 2 • Lesson 9
Page 26

1. 80
2. 57
3. 35
4. 3

Chapter 2 Test
Pages 27-28

1. 97
2. 88
3. 99
4. 76
5. 74
6. 84
7. 91
8. 63
9. 81
10. 0
11. 23
12. 34
13. 7
14. 8
15. 26
16. 63
17. 58 + 21 = 79
18. 48 – 24 = 24
19. 20 – 18 = 2;
    2 + 12 = 14
20. 26 + 38 = 64;
    64 – 20 = 44
21. 36
22. 60
23. 36
24. 25

Chapter 3

Chapter 3 • Lesson 1
Page 29

1. even
2. odd
3. odd
4. even
5. even
6. odd

Chapter 3 • Lesson 2
Page 30

Pairs of objects should be circled for items 1–3.

1. 2, 4, 6, 8
2. 2, 4, 6, 8, 10, 12
3. 2, 4, 6, 8, 10, 12, 14, 16

Chapter 3 • Lesson 3 Page 31

1. 10, even
2. 19, odd
3. 14, even
4. 14, even

Chapter 3 • Lesson 4 Page 32

Groups of five objects should be circled for items 1–3.

1. 5, 10, 15
2. 5, 10, 15, 20
3. 5, 10, 15, 20, 25

Chapter 3 • Lesson 5 Page 33

1. 3 + 3 = 6
2. 15

Chapter 3 • Lesson 6 Page 34

1. 2 + 2 + 2 = 6
2. 5 + 5 = 10

Chapter 3 Test Pages 35-36

1. odd
2. even
3. even
4. odd
5. 19, odd
6. 16, even
7. 2, 4, 6, 8, 10
8. 2, 4, 6, 8, 10, 12, 14
9. 2, 4, 6, 8, 10, 12
10. 5, 10, 15, 20
11. 5, 10, 15, 20, 25, 30
12. 12
13. 15
14. 3 + 3 + 3 + 3 = 12
15. 4 + 4 + 4 = 12

Chapter 4

Chapter 4 • Lesson 1 Page 37

1. 300
2. 500

Chapter 4 • Lesson 2 Page 38

1. 2 hundreds, 3 tens, 5 ones
2. 6 hundreds, 0 tens, 5 ones

Chapter 4 • Lesson 3 Page 39

1. 195, 200, 205, 210, 215
2. 645, 650, 655, 660, 665
3. 530, 535, 540, 545, 550
4. 795, 800, 805, 810, 815

Chapter 4 • Lesson 4 Page 40

1. 220, 230, 240, 250, 260
2. 360, 370, 380, 390, 400
3. 840, 850, 860, 870, 880
4. 930, 940, 950, 960, 970
5. 760, 770, 780, 790, 800

Chapter 4 • Lesson 5 Page 41

1. 500, 600, 700, 800, 900
2. 385, 485, 585, 685, 785
3. 550, 650, 750, 850, 950
4. 420, 520, 620, 720, 820

Chapter 4 • Lesson 6 Page 42

1. 354
2. 420

Chapter 4 • Lesson 7 Page 43

1. three hundred twenty-five
2. eight hundred forty-two
3. six hundred ninety-nine
4. seven hundred thirty
5. one hundred five
6. five hundred seventeen

Chapter 4 • Lesson 8 Page 44

1. 200 + 40 + 1
2. 800 + 0 + 5
3. 200 + 30 + 3
4. 500 + 10 + 2
5. 100 + 10 + 0
6. 400 + 90 + 9

Chapter 4 • Lesson 9 Page 45

1. <
2. =
3. >

4. <
5. >
6. =

Chapter 4 Test
Pages 46-47

1. 2 hundreds, 0 tens, 0 ones
2. 4 hundreds, 5 tens, 6 ones
3. 3 hundreds, 0 tens, 7 ones
4. 445, 450, 455, 460, 465
5. 280, 290, 300, 310, 320
6. 335, 435, 535, 635, 735
7. 528
8. five hundred twenty-eight
9. 500 + 20 + 8
10. >
11. <
12. =
13. <
14. =
15. <
16. >
17. >
18. <
19. >

Chapters 1–4 Review
Pages 48-53

1. 6
2. 14
3. 17
4. 13
5. 4
6. 0
7. 14
8. 5
9. 4 + 9 = 13
10. 8 – 6 = 2; 2 + 7 = 9
11. 15 – 6 = 9
12. 6 + 10 = 16; 16 – 8 = 8

Drawings for items 13 and 14 will vary.

13. 11 + 6 = 17
14. 14 – 8 = 6
15. 49
16. 86
17. 65
18. 74
19. 35
20. 21
21. 46
22. 53
23. 67
24. 35 – 9 = 26 books
25. 67 – 48 = 19 blocks
26. 48 – 20 = 28; 28 + 35 = 63 bikes
27. 83 – 56 = 27; 27 + 30 = 57 nails
28. 28 + 53 = 81; 81 – 47 = 34 stamps
29. 49 + 19 = 63; 63 – 40 = 23 apples
30. 28
31. 68
32. even
33. odd
34. odd
35. even
36. 2, 4, 6, 8, 10
37. 5, 10, 15, 20, 25
38. 15
39. 8
40. 9, odd
41. 14, even
42. 4 + 4 + 4 = 12
43. 2 hundreds, 4 tens, 7 ones
44. 255, 260, 265, 270, 275
45. 320, 330, 340, 350, 360
46. 500, 600, 700, 800, 900
47. 452
48. four hundred fifty-two
49. <
50. >
51. =
52. >

Chapter 5

Chapter 5 • Lesson 1
Page 54

1. 56, 56
2. 89, 89

Chapter 5 • Lesson 2
Page 55

1. 21, 21
2. 67, 67

Chapter 5 • Lesson 2 (continued)

3. 96, 96
4. 72, 72

Chapter 5 • Lesson 3 Page 56

1. 819
2. 880
3. 837
4. 818
5. 593
6. 511
7. 957
8. 589

Chapter 5 • Lesson 4 Page 57

1. 208
2. 449
3. 768
4. 383
5. 59
6. 118
7. 855
8. 100

Chapter 5 • Lesson 5 Page 58

1. 228, 228
2. 275, 275

Chapter 5 • Lesson 6 Page 59

1. 249, 249
2. 826, 826
3. 904, 904
4. 582, 582

Chapter 5 • Lesson 7 Page 60

1. 537
2. 396
3. 968
4. 463
5. 790
6. 754
7. 369
8. 885

Chapter 5 • Lesson 8 Page 61

1. 416
2. 227
3. 169
4. 203

Chapter 5 • Lesson 9 Page 62

Circle should be drawn around answers as follows:

1. add ones
2. 290 + 648
3. Sample answer: He can subtract 437 from 720. The answer is the missing addend. 720 – 437 = 283
4. Sample answer: The number 3 in 300 can be subtracted from the 8 in 873. The answer is 573.
5. The numer 8—the number of tens in 80—can be added to 1—the number of tens in 210. The answer is 290.

Chapter 5 • Lesson 10 Page 63

1. 772
2. 279

Chapter 5 Test Pages 64-65

1. 62, 62
2. 71, 71
3. 78, 78
4. 274, 274
5. 413
6. 277
7. 497
8. 204
9. 88, 88
10. 70, 70
11. 468
12. 771
13. 783
14. 453
15. 118
16. 600
17. 124

Chapter 6

Chapter 6 • Lesson 1 Page 66

1. 8:25 A.M.
2. 7:40 A.M.
3. 2:50 P.M.
4. 12:40 P.M.

Chapter 6 • Lesson 2 Page 67

1. 11:10 on digital clock; 11:10 A.M.
2. 7:50 on digital clock; 7:50 A.M.
3. 4:20 on digital clock; 4:20 P.M.
4. 12:55 on digital clock; 12:55 P.M.

Chapter 6 • Lesson 3 Page 68

1. 6¢
2. 11¢
3. 9¢
4. 4¢

Chapter 6 • Lesson 4 Page 69

1. 20¢
2. 30¢
3. 10¢
4. 25¢

Chapter 6 • Lesson 5 Page 70

1. 30¢
2. 50¢
3. 90¢
4. 60¢

Chapter 6 • Lesson 6 Page 71

1. 34¢
2. 20¢
3. 7¢
4. 24¢

Chapter 6 • Lesson 7 Page 72

1. 75¢
2. $1.50
3. $2.25
4. 100¢, $1.00

Chapter 6 • Lesson 8 Page 73

1. $1.32
2. 85¢
3. 92¢

Chapter 6 • Lesson 9 Page 74

1. 25¢, 28¢, no
2. 79¢, 79¢, yes

Chapter 6 Test Pages 75-76

1. 8:25 A.M.
2. 5:50 P.M.
3. 9:10 on digital clock; 9:10 A.M.
4. 2:35 on digital clock; 2:35 P.M.
5. 25¢
6. 9¢
7. 70¢
8. 37¢
9. 46¢
10. $1.75
11. 70¢
12. $2.19
13. 55¢, 55¢, yes
14. $1.14, $1.19, no

Chapter 7

Chapter 7 • Lesson 1 Page 77

1. 1st and 3rd items should be circled
2. 2nd item should be circled
3. 2nd and 4th items should be colored in

Chapter 7 • Lesson 2 Page 78

1. 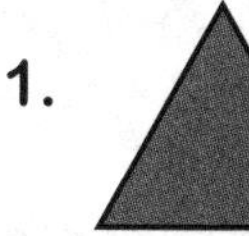
2.

Answer Key

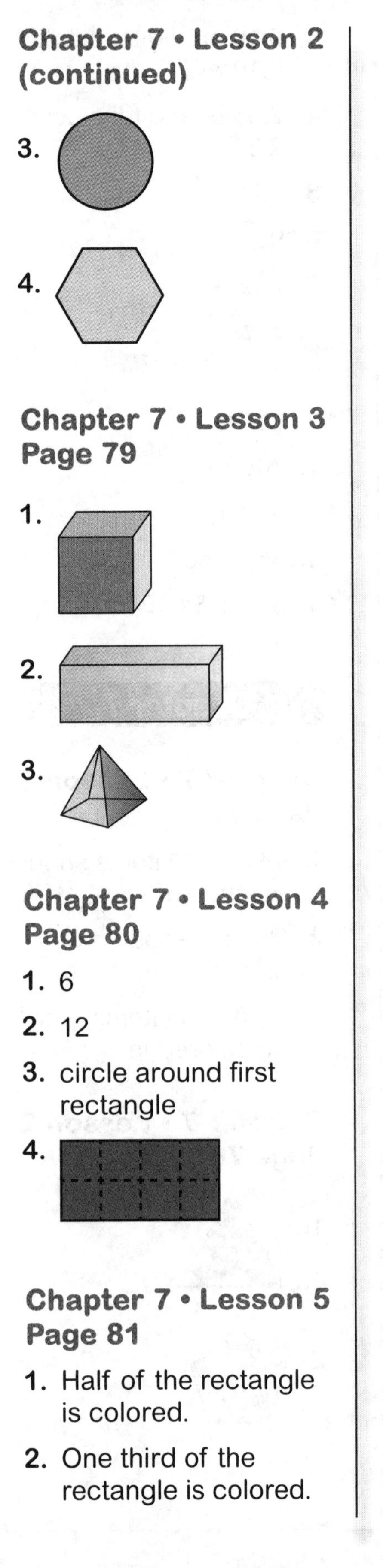

Chapter 7 • Lesson 2 (continued)

3.

4.

Chapter 7 • Lesson 3 Page 79

1.

2.

3.

Chapter 7 • Lesson 4 Page 80

1. 6
2. 12
3. circle around first rectangle
4.

Chapter 7 • Lesson 5 Page 81

1. Half of the rectangle is colored.
2. One third of the rectangle is colored.
3. Lines dividing rectangle into fourths.
4. Bottom rectangle is circled.

Chapter 7 • Lesson 6 Page 82

1. second rectangle is circled
2. 3, or three
3. 2, or two
4. circle around words "three thirds" and "one whole"

Chapter 7 • Lesson 7 Page 83

1.

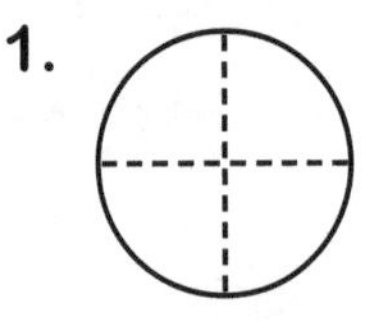

2. last circle should be colored in
3. first circle shows thirds, box around this circle
4.

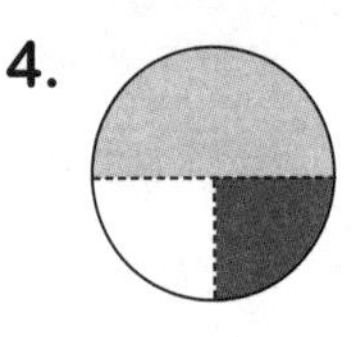

Chapter 7 • Lesson 8 Page 84

1. 2nd circle has box around it, yes
2. third circle shows two fourths and should have a box around it, no
3. words "three thirds" should be circled
4. words "four fourths" should be circled

Chapter 7 • Lesson 9 Page 85

1. First and second circles show half of the circle with color, and should be circled.
2. Possible answers:

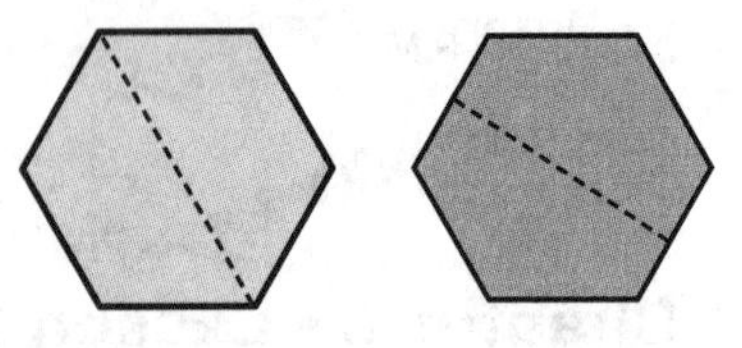

3. Possible answers:

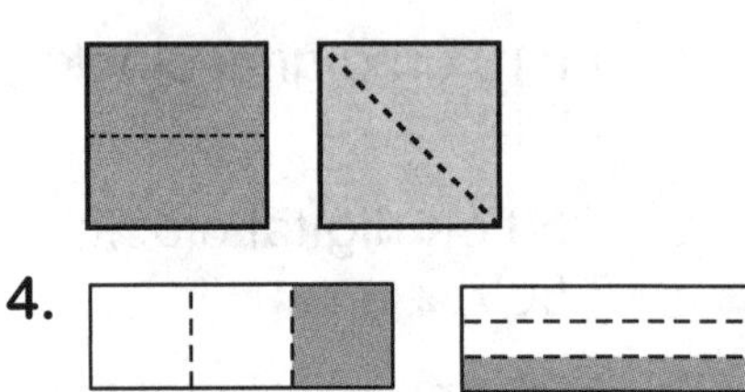

4.

Chapter 7 Test Pages 86-87

1. word "triangle" should be circled
2. word "pentagon" should be circled
3. answers will vary by size, one possible answer is shown

4. blue object is a cube

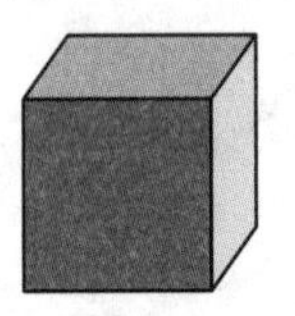

5.

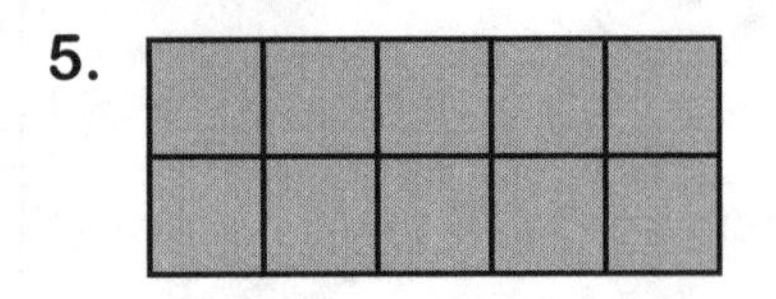

6. Possible answer:

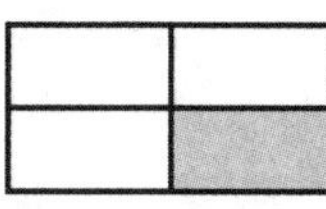

7. "two halves," "one whole;" both circled

8.

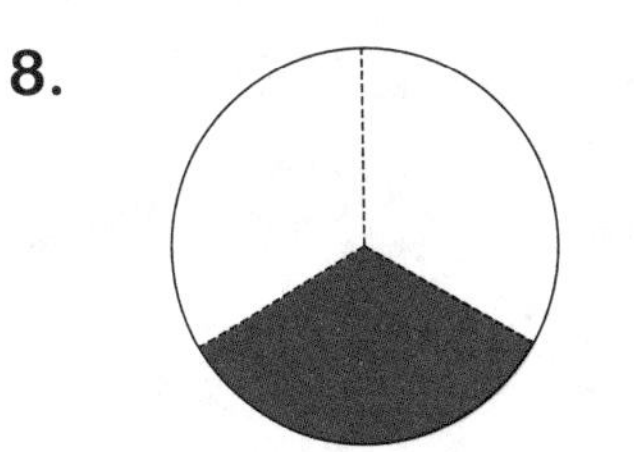

9. words "two halves" should be circled

10. Possible answer:

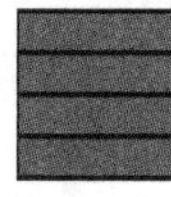

Chapter 8

Chapter 8 • Lesson 1 Page 88

1. 1
2. 2

Chapter 8 • Lesson 2 Page 89

1. 7
2. 10

Chapter 8 • Lesson 3 Page 90

1. 2
2. 2
3. Sample answer: A yard is longer than a foot. It takes fewer yards than feet to measure the same length.

Chapter 8 • Lesson 4 Page 91

1. 400
2. 9
3. 800
4. 2
5. Sample answer: A centimeter is shorter than a meter. It takes more centimeters than meters to measure the same length.

Chapter 8 • Lesson 5 Page 92

1. Estimate will vary. 5

Chapter 8 • Lesson 6 Page 93

1. Estimate will vary. 4

Chapter 8 • Lesson 7 Page 94

1. 2, 1, 1

Chapter 8 • Lesson 8 Page 95

1. 16 – 9 = 7 inches
2. 20 + 12 = 32 feet

Chapter 8 • Lesson 9 Page 96

1.

2. 1

Chapter 8 • Lesson 10 Page 97

1. 17
2. 2

Chapter 8 Test Pages 98-99

1. 2
2. 8
3. 4
4. Inches are shorter so it takes more inches than feet to measure the same length.
5. Estimates will vary. 5
6. 20 − 8 = 12 inches; Greta used 8 inches of string
7. 22 + 21 = 43 meters
8. 15 inches; On the number line, a red dot should be placed over 12
9. 3 fewer miles; On the number line, a blue dot should be placed over 11, and a green dot over 8.

Chapter 9

Chapter 9 • Lesson 1 Page 100

1.

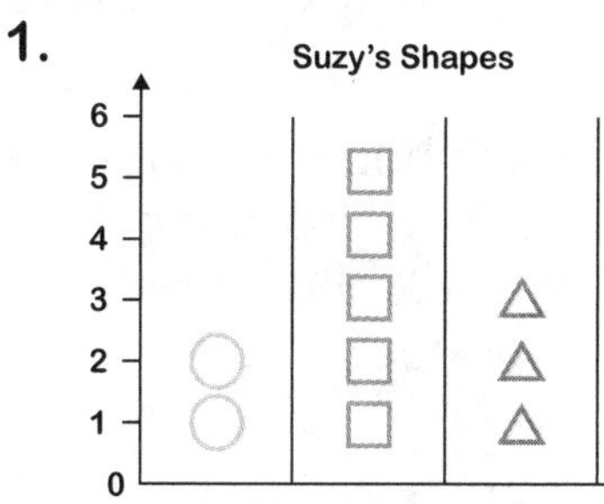

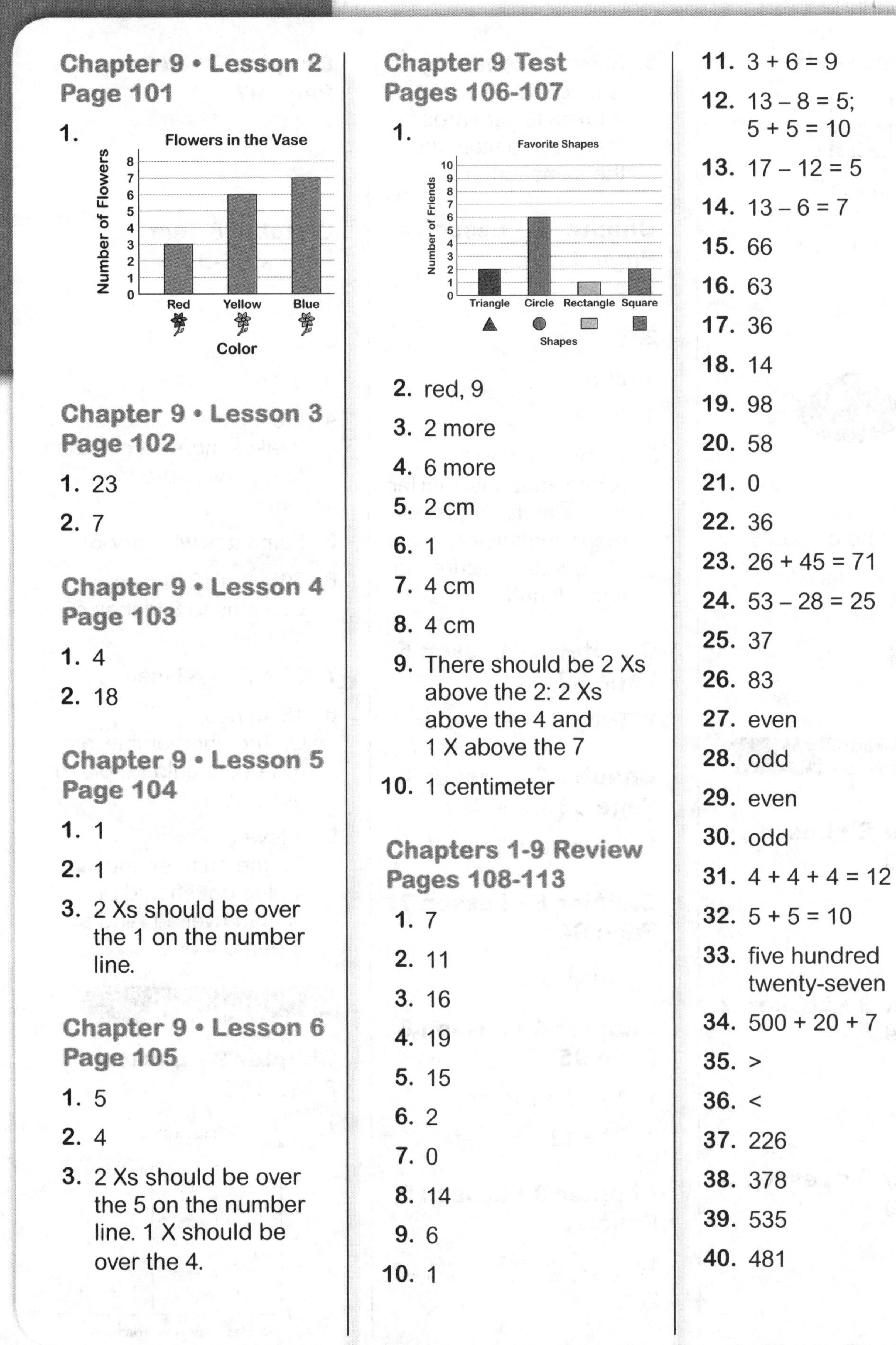

Chapter 9 • Lesson 2 Page 101

1.

Flowers in the Vase

Number of Flowers

0 1 2 3 4 5 6 7 8

Red Yellow Blue

Color

Chapter 9 • Lesson 3 Page 102

1. 23
2. 7

Chapter 9 • Lesson 4 Page 103

1. 4
2. 18

Chapter 9 • Lesson 5 Page 104

1. 1
2. 1
3. 2 Xs should be over the 1 on the number line.

Chapter 9 • Lesson 6 Page 105

1. 5
2. 4
3. 2 Xs should be over the 5 on the number line. 1 X should be over the 4.

Chapter 9 Test Pages 106-107

1.

Favorite Shapes

Number of Friends

0 1 2 3 4 5 6 7 8 9 10

Triangle Circle Rectangle Square

Shapes

2. red, 9
3. 2 more
4. 6 more
5. 2 cm
6. 1
7. 4 cm
8. 4 cm
9. There should be 2 Xs above the 2: 2 Xs above the 4 and 1 X above the 7
10. 1 centimeter

Chapters 1-9 Review Pages 108-113

1. 7
2. 11
3. 16
4. 19
5. 15
6. 2
7. 0
8. 14
9. 6
10. 1
11. 3 + 6 = 9
12. 13 – 8 = 5; 5 + 5 = 10
13. 17 – 12 = 5
14. 13 – 6 = 7
15. 66
16. 63
17. 36
18. 14
19. 98
20. 58
21. 0
22. 36
23. 26 + 45 = 71
24. 53 – 28 = 25
25. 37
26. 83
27. even
28. odd
29. even
30. odd
31. 4 + 4 + 4 = 12
32. 5 + 5 = 10
33. five hundred twenty-seven
34. 500 + 20 + 7
35. >
36. <
37. 226
38. 378
39. 535
40. 481

41. 930

42. 511

43. 906

44. 956

45. Judy can subtract 220 from 538

46. 4:50 P.M.

47. 8:15 A.M.

48. $1.89

49. 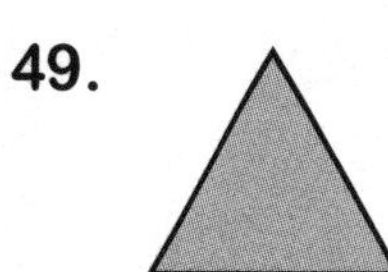

50. The first shape should be circled.

51. The shape in the middle should have a box around it.

52. The second and third shapes should be circled.

53. Sample answer: A foot is shorter than a yard. It takes more feet than yards to measure the same length.

54. Answers will vary.

55. 6

56. 5

57. 12

58. grapefruit juice